LET THEM EAT FLAX

LET THEM EAT FLAX

70 All-New
Commentaries on the Science
of Everyday Food & Life

DR. JOE SCHWARCZ

Director
McGill University Office for Science and Society

ECW PRESS

Published by ECW PRESS
2120 Queen Street East, Suite 200, Toronto, Ontario, Canada M4E 1E2

LIBRARY AND ARCHIVES CANADA CATALOGUING IN PUBLICATION

Schwarcz, Joe
Let them eat flax / Joe Schwarcz.
Includes index.

ISBN 1-55022-698-3

1. Chemistry—Popular works. 2. Chemistry—Miscellanea. I. Title.

Q173.S378 2005 540 C2005-904297-4

Copyedit and Production: Emma McKay
Interior design: Yolande Martel
Interior cartoons: Brian Gable
Cover design: Guylaine Regimbald – SOLO DESIGN
Author photo: Tony Laurinaitis
Printing: Transcontinental
This book is set in Stempel Garamond and Koch Antiqua.

The publication of *Let Them Eat Flax* has been generously supported by the Canada Council, the Ontario Arts Council, and the Government of Canada through the Book Publishing Industry Development Program. Canadā

DISTRIBUTION

CANADA: Jaguar Book Group, 100 Armstrong Avenue,
Georgetown, Ontario L7G 5S4

UNITED STATES: Independent Publishers Group, 814 North Franklin Street,
Chicago, Illinois 60610

PRINTED AND BOUND IN CANADA

ECW PRESS
ecwpress.com

CONTENTS

Introduction   11

Let Them Eat Flax   15
Pomegranate and Blueberry Frenzy   21
"Acrylawhaaat?"   24
Trans Fats   29
Newfangled Chocolates   33
The Sour Side of High-Fructose Sweeteners   36
Cinnamon and Health   39
How Splendid Is Splenda?   42
Spare Me the Wheatgrass Enzymes   45
Functional Foods—From Cod Liver Oil to Vitaballs   49
Eating Shellac   52
Nasty Microbes   55
The French Paradox   59
Jittery Goats and Coffee Beans   62
To Label or Not to Label, that Is the Question!   66
Smoked Meat   69
Nutrigenomics   73
The Saga of Golden Rice   76
Pesticide Problems   80
Organic Agriculture   83

Bringing Piggies to Market   86
Calcium and Weight Loss   90
Learning from the Bushmen of the Kalahari   93
The Cheesecake Factory   96
A Bump for the Antioxidant Bandwagon   100
Vitamin E Doesn't Deliver Either   103
The Cold Facts about Vitamin C   108
A Cancer Treatment on Trial   112
"The Cure for All Cancers"   115
Praying for Health   119
Germs, Germs Everywhere   122
Music and the Savage Breast   125
A Stink about Antiperspirants   128
The Dangers of Betel Beauties and Fruit-Eating Bats   132
Murder by Toxin   135
The Prickly Problem of Acupuncture   139
Reefer Madness   145
Obsessions and Compulsions   150
Water Bottle Confusion   155
A Toxicological Nuance   159
Some Crooked Chemistry   162
The Cryonic Quest for Immortality   165
The Cold Facts About Antifreeze!   168
Would You Like to See My Etchings?   172
Does She or Doesn't She?   175
It May Be an Alcohol—But Don't Drink the Methanol!   181
Citius, Altius, Fortius   185
It Was the Strychnine!   188
Mercury—Pretty but Nasty   191
Feeding the Soil   196

Ammonium Nitrate—More than a Fertilizer   201

Chlorate and the Exploding Trousers   204

The Music of Copper Sulfate   207

Flying High with Aluminum   210

Blue Garlic and Gold Smudge   213

A Stable Mass of Bubbles   216

The Paper Trail   220

The Birth of the Pill   226

The Greatest Inventor   233

Pepper's Ghost   236

Stradivarius or Nagyvaryus?   239

It's Dynamite!   242

The Pox—Both Cow and Small   245

Tin Plague   248

Firebombs, Bedpans, and a Moldy Cantaloupe   252

Radar and Hot Coffee   258

Spontaneous Human Combustion   261

Forceful Sole Searching   265

Magical Hydrides   269

Natural Cures "They" Don't Want You to Know
    About   273

Index   277

INTRODUCTION

There's never a dull moment in the business of interpreting science for the public. Each day seems to bring an onslaught of fresh scientific studies that pertain to virtually every aspect of our life. I look forward to wading through these, but it is increasingly challenging to avoid drowning in the data. Information overload is a vexing problem! For me, the real difficulty lies in trying to distil some practical sense out of the flood of research findings. I can certainly appreciate the journalistic temptation to come up with seductive headlines for stories, but my concern is that often these oversimplify the results of published research, and, in the end, they either provoke unnecessary fears or raise unrealistic hopes. As Mark Twain quipped, "There is something fascinating about science. One gets such wholesale returns of conjecture out of such a trifling investment of fact."

Take, for example, a paper that appeared in the *Journal of Agricultural and Food Chemistry* entitled "Which Polyphenolic Compounds Contribute to the Total Antioxidant Activities of an Apple?" Because apples are dear to our hearts, and "antioxidants" have a positive public image, it wasn't surprising that the press reported widely on the results of the research with headlines such as "Red Delicious Best Disease Fighter." The impact was noted almost immediately with increased sales of

Red Delicious apples. Now, I have nothing against these apples; in fact, I like them. But this study did not show that they fight disease better than other apples! To do that, you would have to follow two groups of subjects for many years, with one group regularly eating Red Delicious apples, the other eating some other variety. What the study did show was that Red Delicious apples have a higher level of antioxidants than some other apples, although varieties such as Jonagold, which is known to be high in antioxidants, were not included. I think we can safely say that fruits are good for us, and that at least part of the benefit likely comes from their antioxidant content. However, it is unrealistic to imply, based on this apple antioxidant study, that substituting Red Delicious apples for others is going to have an impact on overall health. This difference in antioxidant content, relative to the total amount of antioxidants we consume, is not likely to be of practical significance. By all means eat apples—of any variety, along with loads of other fruits and vegetables—but don't assume that Red Delicious apples have some special magical quality. No food does.

So what do we say to breast cancer patients who have read about a study carried out at the University of Texas M. D. Anderson Cancer Center that suggests turmeric, the yellow spice widely used in Indian cuisine, may help stop the spread of breast cancer? First, let's take a look at what the researchers actually did. Based on earlier studies that showed a lower rate of cancer in people who had a diet rich in turmeric, and some previous evidence that one of its ingredients, curcumin, had an anti-tumor effect in the laboratory, the scientists decided to investigate curcumin's anti-cancer potential in a living species. They produced tumors in mice by injecting them with human breast cancer cells, and then surgically removed the cells to mimic a mastectomy. Some of the animals received no further treatment; some were treated with curcumin, some with the cancer drug paclitaxel (Taxol), and others with a combo of curcumin and paclitaxel. The curcumin clearly had an effect: 95 percent of the untreated animals went on to develop lung cancer, but only 50 percent of those treated with curcumin developed tumors. When combined with paclitaxel, the results were even better, with only 22 percent of the mice showing lung tumors. But what does this mean in human terms? Again, it would be unrealistic to suggest that eating curry prevents the spread of cancer. Nobody knows how effectively curcumin is absorbed from the digestive tract, or if it actually has an effect in humans. How much curry would we have to eat? Nobody knows. What we can say is that, based on such studies, it is time to carry out a human trial. Labeling turmeric as an anti-cancer spice is premature and may give false hope.

Putting scientific studies into perspective is now more important than ever because we are on the verge of suffering from health and safety advice overload. As study piles upon study—often with apparently contradictory findings—many people are throwing their arms up in frustration. One study shows that Echinacea may help the common cold; another says it's practically useless. Depending on which study you read, vitamin E is good for almost anything that ails you, or is totally ineffective. In fact, it may even be harmful. Coffee may raise your blood pressure according to one report, while another one finds that it is the number one source of antioxidants in the North American diet. The consequence may be that consumers stop listening to any advice. That's why it is important to emphasize that science is based on a continuous evaluation of all studies until a consensus is reached, and that making lifestyle decisions based on any individual study is rarely warranted. Especially if you believe Dr. John Ioannidis, an epidemiologist at the University of Ioannina School of Medicine in Greece, whose paper in the *Journal of the American Medical Association* claims that there is less than a 50 percent chance that the results of any randomly chosen scientific paper are reliable. His analysis suggests that, due to problems with experimental and statistical methods, small sample sizes, researcher bias, and selective reporting, most research findings cannot be trusted. I suppose this includes his findings as well.

Isaac Asimov, the famed science writer, put it very well when he noted that science now gathers knowledge faster than society gathers wisdom. Let's see what we can do about gathering knowledge and interpreting it with wisdom. And you know what? Eating flax may help us do just that. At least one study claims flax can increase mental prowess. But of course, the study could be wrong.

Let Them Eat Flax

Hippocrates' prescription for his patients who suffered abdominal pains was simple: "Let them eat flax!" And it's probably not bad advice—as long as the pain stems from constipation. It turns out that flaxseeds, which come from the plant used to make linen, are an excellent source of dietary fiber. This indigestible plant component provides a laxative effect by allowing wastes to absorb water as they journey through the digestive tract. But modern science suggests that eating flax may do more than increase the frequency of bathroom visits. How about decreasing the risk of heart disease and cancer? Could Charlemagne really have been on to something when, in the eighth century, he decreed that his subjects should consume flax regularly? It seems so.

Let's begin our story in an unusual place. The barnyard! Not any old barnyard, mind you, but one where the chickens dine on flaxseeds instead of the usual chicken feed. Why? Because some egg producers are trying to improve the nutritional value and the public image of eggs. Let's face it, when "eggs" are mentioned, the first word that often comes to mind is "cholesterol," which in turn conjures up thoughts of clogged arteries and premature demise. In truth, blood cholesterol responds much more to the saturated fats found in meat and full-fat dairy products than it does to cholesterol in egg yolk. Still, eggs suffer from an image problem. Omega-3 fats, on the other hand, positively bask in the limelight these days. Found mostly in fish, these fats have been linked with a reduced risk of heart disease, breast cancer, inflammatory bowel disease, Alzheimer's disease, and arthritis. Slipping these fats into eggs would certainly be a healthy boost to their image! Especially considering that many people worry about pollutants like mercury and PCBS, both of which crop up in fish.

Flaxseed is one of the few plant sources high in omega-3 fats. The term "omega-3" refers to the molecular structure of these fats, indicating the presence of a carbon-carbon double bond on the third carbon from the end of the molecule. Alpha-linolenic acid (ALA), the specific omega-3 found in flaxseed, differs slightly from eicosapentaenoic acid (EPA), and docosahexaenoic acid (DHA), which are the major fats in fish, but some ALA is converted to EPA and DHA in the human body, as well as in the chicken body.

Most research has focused on the health benefits of EPA and DHA, but ALA itself has also been linked with a reduced risk of heart disease. The Nurses' Health Study, monitored by Harvard University researchers, has followed the health status of over 75,000 nurses who, starting in 1984, filled out food questionnaires every four years. Women who consumed the most alpha-linolenic acid from foods had a 46 percent lower risk of sudden cardiac death than women who consumed the least. The major sources of ALA were green leafy vegetables, walnuts, canola oil, and flax.

Canadian guidelines recommend 1.1 grams of omega-3s for women and 1.5 grams for men on a daily basis. These are really no more than educated guesses based on studies carried out

mostly with fish oils. A British trial, for example, showed that heart patients advised to eat two servings of oily fish a week, or to take daily fish oil capsules for two years, had a significantly lower death rate than patients who were told to increase their fiber intake and reduce fat consumption. An Italian study of over 2,800 heart-attack survivors also showed that fish oil capsules providing 850 milligrams each of EPA and DHA dramatically reduced the incidence of death in the first nine months following a heart attack. The protection, however, seems to fade with time, even if fish oil consumption is maintained.

Where do omega-3 eggs fit into this picture? Feeding flax-seeds to chickens makes great use of the "you are what you eat" phenomenon, and results in eggs that have roughly twelve times more omega-3 fats than regular eggs. Of course, the important question is whether eating such eggs makes a significant contribution of omega-3s to the diet. Perhaps surprisingly, it does.

Each egg has roughly 0.35 grams of ALA and 0.13 grams of EPA and DHA, so a couple of eggs provide a significant portion of the recommended intake, about the same as a couple of ounces of a high-oil fish, like salmon. No nutritional authorities suggest that we should be eating two eggs every day, but five to seven a week is reasonable. And even at that rate, switching to omega-3 eggs makes sense. This is roughly equivalent to a weekly serving of fish.

By now you're thinking that this must be too good to be true. There must be a "but" coming up, right? Right! Back in 1994, the scientific community was stunned by a study that linked high blood levels of alpha-linolenic acid with an increased risk of prostate cancer. Total fat consumption had been associated with this cancer before. That was no great surprise, since dietary fat is known to increase the production of male sex hormones, which are linked to prostate cancer. Furthermore, many pesticides are fat soluble, and a high-fat diet increases the body's pesticide load, which is certainly undesirable. But all previous indications had been that a diet high in fish oils decreases the risk of prostate cancer. Could ALA be different from other omega-3s? Was it protecting the heart while increasing the risk for prostate cancer? Several studies since have also suggested that ALA may be linked to prostate cancer, but there is considerable controversy surrounding the issue. Plasma levels of ALA, for example, show no association with ALA levels in tissue taken from prostate cancer patients. The prudent analysis of the data suggests that it is probably not a good idea for men to consume flax oil on a regular basis.

Ready for another "but"? Consuming flaxseeds themselves has consistently been linked to a reduced risk of both prostate cancer and breast cancer. Perhaps this is because other components of the seed, such as lignans, have proven anti-cancer properties, and may overcome any detrimental effect that may be attributed to ALA. A study at Duke University clearly showed that men awaiting surgery for prostate cancer benefited from a daily consumption of three tablespoons of ground flax. Testosterone levels were lowered, and there was a decrease in cancer cell proliferation.

So we now have the following scenario. The omega-3 fat in flax protects against heart disease, probably by reducing inflammation in the arteries and by smoothing out irregular heartbeats. Flax is also an excellent source of soluble fiber, which binds bile acids in the intestinal tract. This forces the liver to make more bile acids to aid in digestion. Since the raw material for bile acid synthesis is cholesterol, flax consumption ends up lowering cholesterol levels. Soluble fiber also slows the transit time of food through the stomach and small intestine, which in turn reduces the rate at which glucose is absorbed into the bloodstream. Diabetics have seen blood glucose levels drop with increased flax intake; in one study, blood glucose levels were lowered by almost 30 percent with a daily consumption of 50 grams of flaxseed. The lignans in the seed might also have a protective effect against cancer. Certainly, the work of Drs. Lillian Thompson and Paul Goss at the University of Toronto is encouraging. These researchers showed that women awaiting surgery for breast cancer had slower-growing tumors if they ate muffins containing 25 grams of milled flaxseed on a daily basis.

Basically, then, consuming ground flaxseed in the ballpark of 25 to 50 grams a day (about 2 tablespoons) seems to be a good idea. Ground seeds can be stored in an airtight container in the refrigerator for roughly a month. Unfortunately, if the seeds are not ground, they tend to exit the body undigested. But consuming flax oil may be a different matter. For men, at least, there is that bothersome potential connection between ALA and prostate cancer.

Omega-3 fats cannot all be lumped into the same category, and it certainly appears that the health benefits of the ones found in fish oil are superior to those of the ALA in flax. Wouldn't it be great if the ratio of ALA to DHA and EPA in flax could be altered to increase the latter? Well, it looks like genetic engineering may just deliver the goods in this instance!

Genetic modification of plants has been criticized for various reasons, including the fact that so far the consumer has seen no obvious direct benefit of the technology. Now researchers at the University of Hamburg have succeeded in modifying flax plants to produce more DHA and EPA. They managed to isolate the gene from a species of algae that codes for an enzyme that converts ALA into DHA and EPA and have introduced it into flax plants. (Fish derive their omega-3 fats from eating algae.) This will make not only for healthier flax for human consumption, but also for improved animal feed. Chickens that dine on genetically modified flaxseeds will produce eggs with a higher DHA and EPA content, and men will worry less about the alpha-linolenic acid content of flax oil. Where does this leave us? Ground flaxseed is a great addition to the diet, and may be even better when the genetically modified version becomes available.

And in addition to all this, omega-3 fats may even enhance brain function. When we are born, our brain already weighs 70 percent of its adult weight, and most of its growth is completed by about six years of age. Infants whose mothers took extra EPA and DHA during pregnancy show higher mental processing scores and eye-hand coordination at age four. Some studies also have shown a beneficial role for these fats during preschool years in terms of preventing attention deficit hyperactivity disorder and enhancing learning capability. So, like Hippocrates said, "Let them eat flax!" Chickens, and people.

POMEGRANATE AND BLUEBERRY FRENZY

"How much pomegranate juice should I drink?" "How many blueberries do I have to eat to get that cholesterol down?" When I get a flurry of such questions, it usually means that a report of a legitimate scientific study has appeared in the lay press, often presenting the results in an overly optimistic light.

The pomegranate craze was sparked by a couple of studies that suggested the fruit may have a role in treating breast cancer and in lowering the risk of heart disease. By the time the tabloids got through with their interpretation of the results, pomegranate juice had become the new wonder kid on the block. And, needless to say, pomegranate capsules are now featured in health-food stores as cancer-preventatives and as treatments for menopause.

But what did the researchers really find? They discovered that there are compounds in pomegranate juice that have estrogenic activity. In other words, they can alter the way that cells respond to the body's own estrogen. This is certainly of great interest because more than two-thirds of breast cancers are estrogen positive, meaning that the body's estrogen stimulates the proliferation of tumor cells. Any substance that reduces such estrogenic stimulation is most welcome. And it seems that some of the polyphenols in pomegranate can do just that. They block the activity of an enzyme known as "aromatase," which is involved in the synthesis of estrogen. (Drugs known as "aromatase inhibitors" are now commonly prescribed in the treatment of some breast cancers.) How did the scientists determine the aromatase blocking activity of pomegranates? By studying the effect of the juice on breast cancer cells in the laboratory. They discovered that extracts of the seeds, which is what pomegranate juice really is, reduced the activity of 17-beta-estradiol, the estrogen of concern in breast cancer, by

some 50 percent. And breast cancer cells that experienced this reduction in estrogen stimulation died with much greater frequency than normal cells. Of course, this is a laboratory finding, and is still a long way away from showing that pomegranate juice has any effect on actual cancers in the body. There is a big difference between bathing cultured cancer cells in pomegranate juice in a petri dish and drinking the juice. Nobody knows if the active ingredients can be absorbed from the digestive tract and if they have any chance of making it to the site of a tumor. But it seems a pretty good bet that pomegranate juice is not harmful, and may do some good.

Although its benefits for breast cancer may be iffy, pomegranate's role as a heart disease preventative is on firmer footing. Israeli researchers investigated the effect of pomegranate juice on LDL cholesterol or, in everyday language, "bad cholesterol." The juice reduced the conversion of LDL into its most damaging form, known as "oxidized LDL." This finding really may be more than a laboratory curiosity. Why? Because the researchers also found that when mice specially bred to develop hardened arteries were given pomegranate juice, the size of the lesions in their arteries was reduced by 44 percent. So, basically, while the hype about pomegranate juice may not be completely justified, there is something to it. A daily glass of 8 ounces just may provide surprising benefits. When ten patients with diagnosed atherosclerosis drank a daily glass of pomegranate juice for a couple of years, their blood pressure dropped by 20 percent, and they also experienced a beneficial reduction in the thickness of their carotid artery walls. These effects were not seen in subjects who consumed a placebo drink free of flavonoids, the pigments in pomegranate juice that are believed to be responsible for desirable effects. So, drink the juice, just don't spill any on your clothes. Pomegranate stains are virtually impossible to get out! Ditto for blueberry stains, which you may also have to deal with if you follow the research in that area.

We've heard before about all the good things blueberries may do for us. Anthocyanins, the pigments responsible for the distinct color of the berries, fall into a category of compounds called antioxidants, and a wealth of research suggests that these are good for us. They may discourage blood clot formation, improve night vision, slow macular degeneration, reduce the risk of cancer, and protect brain cells from aging. So far, it is this anti-aging effect that has captured the imagination of the lay press. While nobody has yet shown that humans who load up on blueberries age more slowly, there have been some intriguing rodent studies. At Tufts University in Boston, a group of elderly rats was put on a blueberry-rich diet, while another group was treated to regular laboratory food. The blueberry-treated rats improved in balance, coordination, and short-term memory.

By the time a rat is nineteen months old (equivalent to about seventy years old for a human), the time it takes them to walk a narrow rod before losing balance drops from thirteen to five seconds. But after eating blueberry extract for eight weeks, the old rats managed to keep their balance for eleven seconds! They also negotiated mazes better! This was the study that the press seized upon, and all of a sudden, blueberries were elevated to the status of a wonder food. And now, with the announcement that pterostilbene (another compound found in blueberries) may reduce cholesterol, the nutritional status of the berries has risen to even loftier heights. The truth is that the study in question was not done on humans, and not even on live animals. It was done in the laboratory, on rat liver cells. The researchers did show that pterostilbene activates a specific receptor on these cells that is linked with reducing cholesterol and triglycerides. But nobody knows if this compound, when ingested, does the same thing in a human liver, or if it even gets there. Nobody knows how many blueberries would have to be eaten to lower blood cholesterol, or indeed if they really can do this.

That doesn't mean such research is to be ignored. My guess—hopefully an educated one—is that blueberries should, as often as possible, be a part of the five to ten servings of fruits and vegetables that experts recommend we consume every day. So I'm ready to raise a glass of pomegranate juice to the researchers who have shown that there just may be something special about blueberries.

"Acrylawhaaat?"

When scientists call a press conference, reporters usually expect a dramatic announcement. They've cloned a sheep. They think they've found a way to produce nuclear fusion in a test tube.

They've completed sequencing the human genome. They've discovered why fewer socks come out of a washing machine than go in. But reporters attending the press conference called by Sweden's National Food Administration in April of 2002 heard nothing of the sort. Researchers at the University of Stockholm, they were told, had discovered acrylamide in potato chips, French fries, and in a variety of other popular foods. "Acrylawhaaat?" the scribes pondered. Most had never before heard of the chemical they would soon help make into a household word. A dirty word!

Acrylamide was a known animal carcinogen, the spokesperson explained. It had been unexpectedly found in a number of common foods, and possibly could account for thousands of cases of human cancer every year. Now he had the reporters' full attention. Chips and French fries, as well as some baked goods, had levels of acrylamide hundreds of times higher than the maximum allowed in drinking water, according to standards set by the World Health Organization. Why should there be any acrylamide in drinking water? Because "polyacrylamide" is commonly used in water treatment to coagulate and trap suspended impurities. While polyacrylamide is harmless, it is always contaminated with trace amounts of the material from which it is made, namely, acrylamide. There is no doubt that acrylamide fed in huge doses to rats can cause a variety of tumors, but health authorities agree that the 1 or 2 micrograms of the stuff that might be ingested daily from water with a maximum allowable concentration of 0.5 parts per billion (ppb) is far too little to have any effect. In other words, the benefits of using polyacrylamide to remove water pollutants greatly outweigh any risk it may introduce.

But the Swedish scientists weren't talking about 0.5 parts per billion, they were talking about French fries that had over 400 ppb, and chips that had as much as 1,200 ppb! Such levels, they suggested, could cause cancer in humans. The acrylamide story made the headlines, causing panic in the supermarket aisles and in the boardrooms of food producers. Was this just another "scare-of-the-day" story, soon to be forgotten, or was it important enough to warrant real changes in our eating habits?

Before coming to grips with that question, let's take a moment to explore how the information about the presence of acrylamide in our food supply came to light in the first place. It all started in 1997, with some paralyzed cows in Sweden. Farmers in the Bjare peninsula began to notice that their cows could not stand up properly. When fish breeders found dead fish by the hundreds in their breeding pools, authorities began to suspect an environmental problem. It turned out that they were right. A tunnel was being built nearby, and it had been plagued by water leaks. To solve the problem, over 1,400 tons of a sealant made with polyacrylamide had been pumped into the cracks in the tunnel. Since scientists had long known that high concentrations of acrylamide could affect the nervous system, the paralyzed cows and dead fish suggested that the chemical had leached out into the water table. Further investigation revealed that it was not only cows and fish that were affected, but also tunnel workers, who complained of feeling numbness in their extremities! As can be expected, this terrified the locals and caused cattle to be slaughtered, milk products to be dumped, and vegetables to be thrown away for fear of acrylamide contamination.

This is when Margareta Tornquist of the University of Stockholm got into the game. She had been asked to investigate the extent to which tunnel workers had been exposed to acrylamide. Blood samples were taken and analyzed for the presence of the chemical. For comparison, Tornquist also looked at samples taken from the general Swedish population. The results were stunning! As expected, the tunnel workers had high blood levels of acrylamide, but so did the others. Where was it coming from? Swedish water did not have unusual levels of acrylamide, so the suspicion turned to the food supply. That's when Tornquist discovered acrylamide in chips, fries, breads, cookies, and crackers. As it turns out, it forms naturally in starchy foods that are fried in fat at a high temperature. When rats were fed such foods, acrylamide was found in their blood at much higher levels than when they were fed boiled foods. A frightening picture began to emerge. A carcinogen, formed in significant amounts in common foods, could end up in the blood and be distributed through the body. According to the Swedish National Food Administration, the world needed to be informed of this risk, so it decided to call a press conference.

But wait a minute here. There is no evidence that acrylamide is a human carcinogen. While it is a well-established neurotoxin, a long-term study of over 8,000 workers who manufacture the substance, and therefore have huge exposures, found no link to cancer. Furthermore, it should be understood that our food supply is filled with natural carcinogens. Aflatoxins in peanuts, ethanol in wine, urethane in sherry, styrene in cinnamon, and heterocyclic aromatic amines in beef bouillon are as carcinogenic to rodents as is acrylamide. But we don't eat isolated ingredients; we eat food. And food has numerous anti-carcinogens as well. Broccoli, onions, soybeans, flaxseed, and apples all contain compounds with decided anti-cancer activity. The bottom line, then, is that there is no scientific justification for the statement that acrylamide in food causes thousands of cases of human cancer. On the other hand, there is plenty of scientific justification to recommend cutting back on fatty, fried foods such as chips and fries, for a variety of reasons. So if fear of acrylamide causes people to do that, they will indeed be better off.

In any case, the food industry has responded to the acrylamide issue by mounting a variety of studies to explore just how acrylamide forms during baking, and how levels can be reduced. It didn't take long to discover that the backbone of the acrylamide molecule comes from an amino acid called asparagine. When heated in the presence of glucose, asparagine undergoes a series of reactions that eventually liberate acrylamide. Food chemists now went to work and found that baking or frying at lower temperatures (below 175°C, or 347°F) significantly reduced acrylamide levels, which could be even further lowered by adjusting recipes or cooking conditions. For example, when sodium hydrogen carbonate (baking soda) is used to replace ammonium hydrogen carbonate as a baking agent in gingerbread, acrylamide concentrations are reduced by more than 60 percent. Blanching potato chips in a dilute acetic acid solution before frying leads to large decreases in acrylamide content. Many such changes have already been instituted, and an expert panel commissioned by the National Toxicology Program of the National Institute of Environmental Health Sciences now estimates that we ingest roughly 0.43 micrograms of acrylamide per kilogram of body weight a day in our diet, which is well below the amounts that cause cancer in laboratory animals.

We can also take some comfort in a recent joint study conducted by the Harvard School of Public Health and the Karolinska Institute in Sweden, which found no link between the consumption of acrylamide and the occurrence of colon, bladder, or kidney cancers. The study's researchers, who reported their results in the *British Journal of Cancer* in 2003, performed what is known as a case-control study. They examined the dietary intake of acrylamide among 987 cancer patients and compared it to that of 538 healthy people to see if they could find a link between the disease and the chemical. No such link was apparent: the cancer patients had consumed no more acrylamide than had the healthy subjects. In fact, they associated higher levels of acrylamide in the diet with a lower, not higher, risk of colon cancer. Still, we are not yet ready to declare acrylamide an anticarcinogen. In all likelihood, foods that contain acrylamide also contain other ingredients, such as fiber, which may offer protection against cancer. An Italian study came up with similar results. An examination of over 7,000 cancer victims showed no evidence of a link to consuming fried or baked potatoes.

The question of a link between breast cancer and acrylamide has also been examined in light of the fact that high doses increase the risk of mammary tumors in rats. A Swedish study, published in the *Journal of the American Medical Association* in 2005, found no evidence of a connection after having followed over 43,000 women with an average age of thirty-nine for eleven years. Based on food frequency questionnaires at the beginning of the study, the women were divided into five categories that reflected their intake of acrylamide. Almost 700 women were eventually diagnosed with breast cancer, but there was no significant difference in the risk of the disease relative to the amount of acrylamide consumed.

Finally, let me call your attention to a paper published in the *American Journal of Clinical Nutrition,* which didn't get nearly as much attention as the acrylamide story. Researchers at Tulane University studied over 9,000 people for roughly twenty-five years and found that those who consumed more than three servings of fruits and vegetables a day had an almost 30 percent lower risk of strokes and heart disease than those who ate less. They didn't call a press conference . . . but should have.

Trans Fats

Really, it all started back in the 1980s. Researchers were surprised to find that Scandinavians, while consuming more saturated fat than Americans, had a lower incidence of coronary heart disease. Consumption of such fats, found in meat, dairy products, palm and coconut oil, is known to drive up blood cholesterol, which in turn is linked to an increased risk of heart disease. So why were Americans at greater risk than the Scandinavians? Well, maybe it had to do with the "partially hydrogenated" fats that American producers were pumping into the food supply.

Saturated fats are composed of chains of carbon atoms that are bonded to as many hydrogen atoms as possible. They are "saturated" with hydrogen. Vegetable oils are mostly "unsaturated," meaning that some of the carbon atoms in their molecules are joined to each other with an extra bond instead of being linked to hydrogen. These carbon-carbon double bonds (described as having a "cis" configuration) impart a bend to the molecule. Treating such unsaturated fats with hydrogen gas in the presence of a nickel catalyst adds hydrogen atoms to some of the carbon-carbon double bonds, resulting in "partially hydrogenated" fats. Since unsaturated fats have not been linked with heart disease, such partially saturated fats were expected to have a better safety profile than the saturated variety. And from a practical viewpoint, hydrogenation reconfigures some of the "cis" bonds to a "trans" form, resulting in molecules with straightened carbon chains, which can then pack together more closely. As a result, liquid oils are converted to solid fats suitable for making margarine or shortening for baked goods. Partially hydrogenated fats are also less likely to go rancid on reaction with oxygen than oils. In other words, partially hydrogenated fats seemed to be the answer to cutting down on saturated fats in the diet. As it turns out, though, things that seem too good to be true usually are.

Martijn Katan at the Agricultural University in Wageningen, the Netherlands, suspected that the higher rate of coronary disease in the US as compared with Scandinavia might have something to do with trans fat consumption. So he enlisted fifty-nine volunteers who, for three consecutive three-week periods, ate diets that varied only in their major fat content. Through one cycle the main fat was oleic acid, a monounsaturated fatty acid found in olive and canola oil, another cycle featured saturated fats, and the third cycle incorporated solid "trans" oleic acid. The results were surprising. Compared with saturated fat, trans fat consumption resulted in higher LDL cholesterol ("bad cholesterol") and lower HDL cholesterol ("good cholesterol"). In fact, the ratio of total cholesterol to HDL cholesterol, a measure of heart disease risk, rose 23 percent on the trans fat diet and only 13 percent on the saturated fat diet. Of course, this was a much higher trans fat intake than the typical North American diet, in which roughly 5 percent of the total calories come from such fats, but the message was clear. There may be a problem with trans fats!

Other studies also cast these fats in an unfavorable light. The famous Nurses' Health Study, administered by Harvard University, has been following over 75,000 nurses for years and has linked foods such as cakes, cookies, white bread, and certain margarines—all major sources of trans fats—with a higher risk of coronary disease. Recently, researchers have also associated trans fats with type II diabetes, breast cancer, sudden cardiac death, asthma, and an increased risk of inflammation. Yes, each of these studies can be, and has been, criticized, and I suspect my analysis will also ruffle some feathers. But the fact is that one can scour the scientific literature and not come up with any benefits for consuming partially hydrogenated fats. So, even if the risks are somewhat exaggerated, there is no harm in avoiding these substances.

And now another facet of trans fats has come to light. Dr. Anne-Charlotte Granholm at the Medical University of South

Carolina found that trans fats might impair learning and memory! She fed one group of rats a diet that contained 10 percent hydrogenated coconut oil, a common trans fat, while another group dined on soybean oil. Then the animals had to sink or swim—literally. They had previously been trained to find the location of hidden platforms in a water-filled maze, and now they were asked to recall their training. Well, basically, the soybean rats had no problem, but the trans fat rats floundered. This was not totally unexpected, since previous studies had shown some memory impairment with increased consumption of fats, but now the finger seemed to be directly pointed at trans fats. And the scary thing was that the quantity of trans fats the rats consumed was proportional to what North Americans eat routinely. How trans fats damage the brain isn't clear, but one theory is that they cause inflammation that damages specific proteins that nerve cells use to send and receive signals. Dr. Granholm was so impressed by the results of her research that she went home and rid her kitchen of all foods that harbored trans fats. She also swore off French fries, which are usually loaded with the nasty stuff.

Manufacturers are heeding the advice coming from researchers and are trying to reduce the trans fat content of their products. Trans fat–free Oreos are already on the market, alongside trans fat–free Doritos. A move in the right direction, I suppose. And maybe McDonald's will make good on its promise to eliminate trans fats. But you know what? Foods that contain trans fats are generally poor on the nutrition scale anyway, so there is no harm in cutting down. As far as snacks go, well, apples don't have trans fats. Nor do oranges. Or bananas. Or broccoli. Munch on those instead of doughnuts and you'll be healthier. And, apparently, smarter.

NEWFANGLED CHOCOLATES

When I was growing up in Hungary, my big treat was a cup of hot cocoa every night. Cocoa powder was a luxury item, and I was only able to partake of the pleasure because my aunt, who lived in Canada, used to send us "care" packages. And you know what? I may just go back to that old habit. Especially if the research about "high-flavonol" cocoa turns out to be as promising as it now seems. Dr. Norman Hollenberg of Harvard Medical School certainly thinks it will. And he should know. Hollenberg is involved in serious research about the possible health benefits of cocoa, all because a while ago he came across a scientific paper written back in the 1940s about the unusual blood pressure of the Kuna Indians living in the San Blas Islands of Panama. What was so unusual about the blood pressure of these natives? It was extremely low, and did not rise with age. Did they possess some marvelous gene, Hollenberg wondered? As it turned out, no. Kuna Indians who had moved to the mainland of Panama did not have unusually low blood pressure. So what were they doing on the island that could have had such an amazing effect? It seems they were drinking a lot of cocoa made with locally grown, minimally processed cocoa beans. This, Hollenberg thought, was worthy of investigation.

So where do you go to seek research funds for such a study? You don't go to a lightbulb manufacturer or an automobile producer. You go to the chocolate industry. And the Mars Company, as you can imagine, was quick to jump on this bandwagon. The giant private corporation, which floods the world with the likes of Dove bars and M&Ms to the tune of some $17 billion in annual sales, could well afford to support some promising chocolate research. In fact, the company itself had long been tinkering with the chemistry of chocolate, hoping to come up with a version that could be promoted as having some sort of health benefit.

Mars' work had been stimulated by the fact that chocolate contains flavonols, compounds also present in tea and red wine that have been linked with protection against heart disease. The company had been trying to come up with a high-flavonol cocoa, but had run into difficulty because flavonols impart a bitter taste. Finally, though, after years of work, Mars scientists managed to find the right kind of cocoa beans, which through a patented mild process could be transformed into a high-flavonol cocoa powder acceptable to the palate. Mars was happy to provide Dr. Hollenberg with a supply for his studies. And what did the studies show? That the high-flavonol cocoa relaxed blood vessels in people and led to better circulation. Amazingly, it resulted in a 33 percent improvement of blood flow to the brain, which could be very meaningful for people suffering from dementia caused by poor circulation.

There was other good news about chocolate coming from research supported by Mars at the University of California in Davis. Here, Dr. Carl Keen, chair of the Department of Nutrition, had found that a flavonol-rich cocoa drink had an effect on blood platelets that was similar to taking a daily dose of a baby aspirin. Such "blood thinning" has been shown to reduce the risk of heart attacks caused by blood clots. Mars scientists now toted up the results of their own research, and of work carried out by university investigators, and decided that the time had come to tackle the challenge of making chocolate into a "functional food." Most everyone loves chocolate because of its flavor, but just imagine the market potential if there were significant health benefits to munching on the delicacy. Health instead of guilt? Sounds great!

Enter the "CocoaVia" bar, which Mars is test-marketing on the Internet. Each bar contains a standardized amount of 100

milligrams of flavonols, meaning that the daily consumption of two such bars provides a flavonol content that can have an appreciable effect on blood pressure and platelet aggregation. But the researchers have taken an extra step. They have incorporated a gram and a half of phytosterols into each bar. These plant-derived compounds have been shown to lower blood cholesterol levels. What about the fat content of these bars? Doesn't that outweigh any benefit? Actually, it seems not. Each bar has only about 3 grams of fat and 90 calories. Of course, fruits and vegetables are still a far better source of flavonols and phytosterols, and nobody, including Mars, is suggesting that CocoaVia is a "medicine." But, hey, if scientific research can produce a healthier chocolate, then why not? And if the mechanism of the action of flavonols is unraveled, such research may eventually even lead to flavonol-based medications.

Dr. Hollenberg at Harvard is very interested in determining just how flavonols affect blood pressure. His research has shown that these compounds somehow increase the release of nitric oxide, a substance that causes blood vessels to dilate, in the body. Dilated blood vessels in turn lead to decreased blood pressure. In fact, Viagra works via the same mechanism. It opens up blood vessels, allowing more blood flow to some important anatomical parts. So now you see that there may be several reasons for going back to that childhood regimen of drinking a cup of hot chocolate every night. Dr. Denise O'Shaugnessy, a researcher in England, would likely agree. In 2005, her research team found that just one cup of cocoa can inhibit the functioning of platelets, cells that are involved in blood clot formation. Since blood clots can cause heart disease and strokes, a nice cup of hot cocoa may be just what the doctor orders. And with huge profits at stake, we probably won't have to wait too long until Mars' patented high-flavonol cocoa becomes commercially available. And can healthier M&Ms and Dove bars be far behind? Sweet dreams, but just don't forget about the real proven "functional foods": your fruits and veggies.

THE SOUR SIDE OF HIGH-FRUCTOSE SWEETENERS

Sugar producers are hopping mad. So are the companies that flood soft drinks, cereals, yogurts, baked goods, and various desserts with high fructose corn syrup (HFCS). There is no sweet talk at all when it comes to their reaction to a recommendation by the World Health Organization that the intake of added sugars in food and drink should be no more than 10 percent of daily calories. Sweeteners are being unfairly singled out as being responsible for poor diets, the industry claims, since the cause of obesity is too many calories, no matter where they come from. "Taxpayers' dollars should not be used to support misguided, non-science based reports, which do not add to the health and well-being of Americans, much less to the rest of the world," says the Sugar Association. Humbug, I say. We consume way too much sugar and other caloric sweeteners, and there is plenty of scientific information to suggest a link with obesity and other health problems. Of course the industry objects to such allegations—after all, billions of dollars are at stake.

No, sugar is not "white poison," as some would have us believe. In moderate amounts, it can be part of a healthy diet. But North Americans are not consuming sweeteners in moderate amounts. We are guzzling them at a rate of about 50 teaspoons of added sugar per day! That is an astounding amount. It's readily believable, though, given that a can of pop has roughly 10 spoonfuls, and many people drink several cans a day. In this context, "sugar" refers to both sucrose, which is extracted from sugar cane or sugar beets, and "high fructose corn syrup," which is manufactured from cornstarch. Sucrose consumption has actually declined in the last twenty years, but that's only because it has been replaced by HFCS as the prime sweetener. The increased use of HFCS mirrors the increase in obesity in North America. Of course, such an association cannot prove cause and effect, as I'm sure industry spokespeople would be quick to point out. Overconsumption of high-calorie foods and lack of exercise is the problem, they say. Technically, they are right. But the fact is that far too many of those extra calories come from added sugar. So why not reduce this? There is no downside to curbing our intake of sweeteners. There is absolutely no risk in drinking water instead of soft drinks!

Replacing sucrose with high fructose corn syrup may not be just a benign switch of one caloric sweetener for another. There may be metabolic consequences. So why was this switch made in the first place? Because HFCS costs a few pennies per kilo less than sugar does to produce. But because of the volumes used, that can translate into hundreds of millions of dollars in the long run.

The technology to produce HFCS emerged in the 1950s with the isolation of enzymes from bacteria capable of breaking cornstarch down into glucose. Since the US government subsidizes corn production, a cheap way of producing glucose now became available. There was a problem, though. Glucose is only about 70 percent as sweet as sucrose. Once more, some newly isolated enzymes entered the picture. Glucose isomerase, from a special strain of *Streptomyces murinus*, readily converted some of the glucose into fructose, which is 30 percent sweeter than sucrose. It is also more water soluble than glucose. This made it possible to produce a stable syrup with roughly 55 percent fructose content and high sweetening power. This "high fructose corn syrup" was easier to blend into soft drinks and foods than sucrose, and was welcomed by everyone—except, of course, the sucrose producers.

When HFCS was first introduced, nobody thought it would have a different effect on the body than sucrose. After all, sucrose is broken down in the body to equal amounts of glucose and fructose, and can therefore be thought of as a 50 percent fructose product. Can an extra 5 percent of fructose in HFCS make a metabolic difference? Yes, some researchers argue. Our consumption of fructose has increased by some 30 percent in the last thirty years, and this may have some consequences on obesity. The digestion, absorption, and metabolism of fructose differ from glucose. Fructose does not trigger insulin release, which at first may seem like a good thing. But it might not be so good. Insulin stimulates the production of a hormone called leptin, which inhibits food intake. With less leptin production, food consumption goes up. Leptin also acts on the stomach to prevent the release of ghrelin, the major hormone responsible for hunger. If there is an excess of fructose in the bloodstream, leptin is not increased, and the stomach cells are not stopped from producing ghrelin. We feel hungry, and we eat more. Furthermore, glucose itself provides satiety signals to the brain, but the transporter molecule that fructose uses to enter cells is absent from the brain. And if that isn't enough, fructose is more readily converted into fat inside cells than glucose is.

I know what you're thinking. Fructose is the sugar found in fruits, and everybody knows that fruit is good for us. You're right! But an apple has only about 10 grams of fructose, whereas a 12-ounce can of pop has 25. And because the apple contains fiber, its fructose content is absorbed into the bloodstream much more slowly, resulting in smaller effects on metabolism. But most importantly, fruit is full of antioxidants and minerals conducive to good health. Soft drinks have no such redeeming features. No matter what the sweetener industries say, cutting down on sucrose and HFCS-sweetened foods and drinks will have an impact on the obesity epidemic and will lead to better health. So I say, let's drink to the stand taken by the World Health Organization. Just don't make it a soft drink.

CINNAMON AND HEALTH

Just mention cinnamon, and I can smell and almost taste my mother's apple strudel. She made it from scratch, gently pulling the pastry on a moist tablecloth until the dough was paper-thin. The filling was made with fresh apples and, of course, a liberal sprinkling of cinnamon. Who would have ever thought that this spice added more than just flavor? Like a dose of good health? That's a distinct possibility if we go by a recent study carried out jointly by NWFP Agricultural University in Pakistan and the US Human Nutrition Research Center, which showed that this brown powder from the inner bark of a type of evergreen grown in Asia can help reduce blood levels of glucose, triglycerides, and cholesterol. Who could ask for anything more?

The effect on glucose levels is particularly important because type II diabetes is reaching epidemic proportions in North America. It is being diagnosed at younger ages than ever, mostly due to increasing obesity. In this type of diabetes, cells become less sensitive to insulin, the hormone secreted by the pancreas to stimulate glucose absorption into cells from the blood to be used for producing energy. If this process is impaired, extra glucose in the blood can increase the risk of heart disease, as well as kidney, eye, and circulatory problems. Over the years, a number of foods and beverages have been examined for possible blood glucose–lowering effects. Green tea (but not black tea) has such an effect, as does coffee. The catch with coffee is that you have to drink at least six cups a day, which is certainly not advisable. Researchers are, however, focusing in on the active ingredient, which is not caffeine, but chlorogenic acid. Eventually, we may see this in a pill form. Red wine, probably due to its resveratrol content, also lowers blood glucose at a dose of about three glasses a day, which may be a tad much. Other plant-derived materials have been claimed to reduce blood glucose levels, but the supporting evidence has been sketchy. Fenugreek, bitter melon, Korean ginseng, Gymnema, onions, flaxseed, and cinnamon have been repeatedly mentioned in medical lore. Now researchers have decided to put cinnamon to the test. And it was all because of some apple pie!

Richard Anderson and his colleagues at the Human Nutrition Research Center in Beltsville, MD, were studying the effects of a low-chromium diet on blood glucose levels. Chromium is necessary for proper insulin function, probably by activating an enzyme known as insulin receptor kinase. This enzyme primes certain proteins in cells to act as receptors for insulin. Without chromium, insulin can't bind to these receptors and carry out its function as a "gatekeeper" for the entry of glucose into cells. Our modern fast-food diet is low in chromium, and furthermore, foods high in sugar stimulate elimination of chromium from the body. Some research has actually shown that type II diabetics who have a low-chromium diet can achieve better sugar control with chromium supplements. This has been somewhat controversial because there is no agreement on which kind of supplement is the best; chromium picolinate— which yields readily absorbable chromium—has been used, but it does have a cloud hanging over it. Some studies have suggested that it can disrupt DNA, which is certainly an undesirable side effect. Chromium histidine or "glucose tolerance factor" chromium may be better choices.

In any case, Anderson was interested in studying the effects of a low-chromium, high-sugar diet on type II diabetics. And what food was deemed to be ideal for this study? Apple pie! Virtually no chromium, and lots of sugar. It should have sent blood glucose soaring. But it didn't! The only reasonable explanation for this seemed to be the presence of cinnamon in the pie. After all, there had been all sorts of folkloric evidence for the anti-diabetic properties of cinnamon. Now it was time to see if folklore could be converted to science.

Sixty people with type II diabetes, average age of fifty-two, were enlisted and divided into groups that would consume 1 gram, 3 grams, or 6 grams of powdered cinnamon in capsules after their daily meals. A control group was given capsules with wheat flour, an inert substance. The experiment lasted for forty days.

After forty days, blood glucose was significantly lower in the cinnamon group, in some cases by as much as 30 percent. Interestingly, the people who consumed only 1 gram did as well as those on the higher doses. Total cholesterol, LDL (the notorious "bad cholesterol"), and triglycerides were also significantly reduced. The researchers noted that blood glucose levels stayed low even twenty days after the cinnamon ingestion was stopped, suggesting that it need not be consumed every day for a blood glucose–lowering effect to be observed. Best of all, there seems to be no downside to eating a gram of cinnamon a day. It contributes virtually no calories and even tastes good. And, of course, it is not only diabetics who can avail themselves of the cinnamon advantage. Anyone with high cholesterol can give a gram of cinnamon a day a shot. It isn't hard to incorporate this much into the diet. You can sprinkle some into your coffee, mix some into your cereal, or even make a tea by boiling a stick of cinnamon in water. But I would suggest that apple strudel, no matter how good it tastes, is not the way to go. The butter in it—if it is properly made—outweighs any benefit the cinnamon may have. Ah . . . but it does make my mouth water. . . .

How Splendid Is Splenda?

They tortured that poor molecule. They heated it, froze it, dissolved it in acid, baked it into cakes, stuffed it into people's mouths, and fed it to rodents. They even made it radioactive so they could follow its path through the animals' bodies. Then they extracted it from the rodents' poop to see how it had fared. And it fared well. In fact, it passed every indignity with flying colors. As a result, we now have an artificial sweetener on the market that promises to sugarcoat our lives—but without actually using sugar. Sucralose (Splenda) is ready to challenge aspartame as the leader in the artificial sweetener sweepstakes.

We love sweets. There's no doubt about that. Our palates lust for ice cream, our mouths water at the thought of glazed doughnuts, our parched throats yearn for soft drinks, and visions of sugarplums sometimes dance in our heads. To satisfy these cravings, we gorge on caloric sweeteners such as sugar or high fructose corn syrup. This is not exactly ideal for our teeth, our waistline, or our general health. Thus, it isn't surprising that artificial sweeteners and artificially sweetened products have attracted greater and greater consumer interest.

Unfortunately, none of the sugar replacements has been completely satisfactory. Aspartame, or "Equal," is about 180 times sweeter than sugar but is not stable in acidic conditions or when exposed to heat. This presents significant problems for diet drink manufacture, and for baking. Saccharin and the cyclamates are sweet enough, but leave an aftertaste. There have also been some lingering health concerns. Aspartame cannot be consumed by individuals with an inherited condition called phenylketonurea (PKU), because their bodies lack the ability to metabolize phenylalanine, one of this sweetener's breakdown products. In rare cases, people may have adverse reactions to aspartame, including headaches and visual disturbances. Saccharin and the cyclamates, in turn, have had to cope with the shadow of cancer. Some studies, probably of no relevance to humans, have suggested a slightly elevated risk when test animals were fed massive amounts of these sweeteners. In any case, the nutritional world was primed for a new kid on the block.

The kid was born in a laboratory at Queen Elizabeth College, University of London, in 1976. The researchers studying the chemical reactions of ordinary table sugar, or sucrose, certainly did not have artificial sweeteners on their minds. But when they managed to incorporate three chlorine atoms into a sucrose molecule, they aroused a sugar company's interest. A company representative called one of the researchers to ask for a sample to be tested. As luck would have it, the young foreign chemist misunderstood and thought the request was for a sample to be "tasted." So with a bit of bravado he plopped some of the chlorinated sugar into his mouth and told his supervisor about the sweet experience. The supervisor immediately recognized the potential value of this discovery, seized the moment, and redirected the laboratory's research efforts. Sucralose, as the new compound came to be called, turned out to be 600 to 1,000 times sweeter than sugar, depending on what it was added to!

But many years of testing faced the energized chemists before the new product could be brought to market. Their enthusiasm increased when sucralose turned out to be very water soluble as well as stable to heat and acid. This meant that it could easily be used in diet drinks and baked goods. Since sucralose is so sweet, much less of it is needed than if sugar is used. But this presents a problem. Sugar provides not only sweetness, but also bulk in bakery products. However, when sucralose is combined with maltodextrin, a type of starch that provides bulk, the mixture can be substituted for sugar, measure for measure. The end product isn't necessarily identical, however. Since sugar is also responsible for the browning effect produced by baking, the color of some cookies, for example, may look rather anemic. One thing sucralose is not good for is making fudge! It comes out much too syrupy.

Safety testing of sucralose has been extensive. For fifteen years, it was subjected to a battery of short-term and long-term animal feeding studies. The results were conclusive. Most of the sucralose dose was excreted unchanged, and even the small percentage that was metabolized yielded compounds that were also excreted. Every bit the animals were fed could be accounted for in their excreta. Any concerns about storage in the body or interference with metabolic pathways essentially evaporated. So diabetics can safely use sucralose. As an added benefit, unlike sugar, this sweetener has no detrimental effect on the teeth. While our bodies cannot break down sucralose, microorganisms in water and the soil readily do so. In other words, the stuff is biodegradable and poses no environmental hazard.

Is there anything, then, to criticize about sucralose? We might be wary of some of the marketing approaches that have trumpeted the safety of this compound by referring to the fact that it is made from "natural sugar." What a substance is made from is irrelevant; what matters is what the final product is. Its properties are determined not by its ancestry, but by its molecular structure. Hydrogen gas, for example, can be made from water, but it would be absurd to suggest that it therefore has the same safety profile. It's a different substance, just like sucralose is different from sugar. Incorporation of three chlorine atoms into the sugar molecule converts it into a totally new substance. Sucralose is safe because it has been extensively tested, not because it is made from sugar! One of its other attributes is that it leaves no bitter aftertaste, but unfortunately, I cannot say the same thing for some of the advertising hype about the product.

Spare Me the Wheatgrass Enzymes

"Just try it once, please, just try it," the lady begged me. "OK," I finally said, hoping to bring the discussion to an end. She opened the Thermos bottle she had been clutching and poured me a glass of a green liquid, assuring me that she had squeezed the wheatgrass barely an hour ago. I could therefore be confident, she said, that the enzymes in it were still alive! Well, dead or alive, they certainly did nothing for the taste of the beverage.

This wheatgrass juice was one of the foulest things I've ever tasted. Of course, I was quickly assured that I was not drinking it for taste; I was drinking it for health.

This gustatory calamity followed on the heels of an hour-or-so-long discussion on the merits of consuming chlorophyll and live enzymes. My guest had sought an appointment to open my eyes to a form of therapy that would help millions of people who were being poisoned by eating "dead food." And so it was that I came to learn about the Hippocrates Health Institute and the teachings of Ann Wigmore.

Ann was a Lithuanian émigré to the US who had become convinced of the healing power of grasses after reading the Biblical story of Nebuchadnezzar, the Babylonian king who apparently cured himself of a seven-year period of insanity by eating grass. Wigmore reflected on this story, considered how dogs and cats sometimes eat grass when they feel ill, and came up with a theory about the magical properties of wheatgrass juice. Food rots in the intestine due to improper digestion, she maintained, and forms "toxins" that then enter the circulatory system. The living enzymes in raw wheatgrass prevent these toxins from forming and ward off disease. By 1988 Wigmore, who had no recognized scientific education, was even suggesting that her "energy enzyme soup" was capable of curing AIDS.

Ann Wigmore died in 1994, but the "live enzyme" theory lives on. Numerous books tout the benefits of ingesting enzymes, health food stores stock bottles of enzyme capsules and powders, and restaurants that guarantee low-temperature cooking to prevent the murder of enzymes are sprouting up. No need to worry about killing enzymes, though. They were never alive in the first place. Enzymes are not composed of cellular units; they cannot reproduce, they cannot carry on metabolism, and they cannot grow. Ergo, they are not alive.

"There would be no life without enzymes," begins the usual sales pitch. Well, you can't argue with that statement. Indeed, enzymes are special protein molecules that are involved as catalysts in virtually every chemical reaction that takes place in the body. "Heat can destroy enzymes," the pitch continues, "so processed or cooked food is devoid of these life-giving substances." This is also true. The inference then is that we should be eating "live food" because that's the only way we can get the "live enzymes" our body needs. In the case of Ann Wigmore, it is more than an inference. Her book states: "Each of us is given a limited supply of enzyme energy at birth. This has to last a lifetime. The faster you use up the supply, the shorter your life. Cooking food, processing it with chemicals, using medicines, uses up the enzymes. The Hippocrates Diet makes enzyme deposits into the account." Absurd! Our body does not need ingested enzymes, and, except for specific rare instances, it cannot use them.

Enzymes are proteins, and, like other proteins, they are broken down during digestion. The fact that studies have shown that some enzymes may escape digestion and enter the bloodstream should not be interpreted as a benefit. Enzymes are remarkably specific in their actions, and the enzymes that may make it into the bloodstream from food are not the same as the body's enzymes. Many promoters of "live food" diets emphasize that the "living enzymes" in fresh fruits and vegetables help digestion and spare the body's enzyme supply from being wasted on digestion. The spared enzymes are then said to be free to take part in metabolism and disease fighting. Nonsense. Metabolic enzymes have nothing to do with digestive enzymes. Even if enzymes in raw fruits and vegetables survived passage through the highly acidic environment of the stomach and managed to enhance digestion in the small intestine, they would have no effect on the enzymes involved in the cellular processes that go on all over the body.

This is not to say that oral enzyme therapy is always without merit. People who are lactose intolerant can benefit from ingestion of the enzyme lactase, which is lacking in their digestive tract. But the lactase pills have to be specially formulated to enhance passage through the stomach. Cystic fibrosis patients have to compensate for a lack of pancreatic enzymes by swallowing pills, which are enterically coated to ensure they reach the small intestine. In some cases, people with digestive problems may benefit from plant-derived enzyme supplements that help break down proteins, fats, and carbohydrates. While these enzymes do the job in the laboratory, their effectiveness in the digestive tract is controversial. Researchers are also investigating whether certain oral enzymes may be of use in cancer treatment, but unfortunately, so far, the results haven't been particularly encouraging.

There is yet another bizarre feature of the raw food diet championed by Ann Wigmore and others. They point out that the molecular structure of chlorophyll, the green coloring in plants, is almost the same as hemoglobin, which carries oxygen around the body. They infer that ingesting chlorophyll enhances energy by increasing oxygen transport. This is pure twaddle. Humans are not plants; we do not photosynthesize, and we have no requirement for chlorophyll. In any case, chlorophyll cannot be absorbed.

Now that I've gotten all that off my chest, I'll go on record as recommending a "live food" diet. The fruits and vegetables that make up such a diet contain all sorts of substances that enhance health. But enzymes are not among them. As I related all of this to my office guest, I had a glimmer of hope when she accepted my explanation that oral enzymes in food are unlikely to survive digestion. The apparent victory, though, was short-lived. "Maybe that's why Ann Wigmore was so high on wheat-grass juice enemas," she retorted. Mercifully, she didn't ask me to try one.

Functional Foods—From Cod Liver Oil to Vitaballs

Not all childhood memories are happy ones. I still gag at the thought of swallowing a mixture of raw egg yolk, cod liver oil, and sugar. Of course, I never realized at the time that my mother wasn't trying to torture me; she was practicing what we would now call "functional food chemistry." In other words, she had concocted a food that would deliver something more than simple nutrition. The cod liver oil was an excellent source of vitamins A and D and also, as we have since learned, healthy omega-3 fats.

"Functional foods" have now become big business. Foods and drinks fortified with everything from calcium and fiber to live bacteria and herbs vie for the consumer's attention with promises of enhanced vigor and health. Some, like ginkgo biloba–laced herbal snacks for "improved memory" are just nonsense. Forget about them. Others, such as orange juice with added calcium, fruit drinks with lutein, or omega-3-fortified cheese sauce make sense. In fact, the Texas school system has already begun to serve omega-3-fortified breakfast tacos and cheese sauce to students with hopes of reducing learning disabilities and behavioral problems linked to poor nutrition.

Research in the area of functional foods is exploding. We can, for example, look forward to milk shakes, soups, or puddings designed to control blood pressure. Sounds far-fetched? Actually, it isn't. Researchers at Davisco Foods in the US have been looking at substances derived from whey protein that can reduce blood pressure. Whey, the stuff left behind when cheese is made, harbors a variety of proteins. When these are broken down, or "hydrolyzed," by special enzymes, they yield fragments called peptides. In laboratory experiments, some of these inhibit the activity of angiotensin-converting enzyme (ACE), which plays an important role in blood pressure regulation. Many anti-hypertensive drugs, known as the ACE inhibitors, work by interfering with the activity of this enzyme.

Of course, ACE inhibition in a test-tube experiment does not necessarily mean that the same effect will be seen in living systems. But the researchers were encouraged when the whey peptide, christened BioZate, reduced blood pressure in rats. (Must be tough to apply those tiny blood pressure cuffs.) This launched a human experiment in which thirty men and women with borderline high blood pressure were divided into two groups, one taking the whey protein isolate on a daily basis, and the other taking regular whey protein. After just one week, there was a significant drop in blood pressure in the experimental group (11 mm Hg in the systolic pressure and 7 mm Hg in the diastolic) as compared with the control group. American scientists are not the only ones to have found this effect. Indeed, in Iceland, a beverage containing a milk-derived tripeptide is already available on the market in apple-pear, cherry, or strawberry flavors. LH, as it is called (after the *Lactobacillus helveticus* bacteria used in the process of breaking down the proteins), is marketed in 100 milliliter daily dose bottles to help control blood pressure.

A drink with whey protein to control blood pressure sounds pretty palatable. But how about a worm-laced beverage to treat ulcerative colitis, or Crohn's disease? These "inflammatory bowel diseases" (IBD) have been increasing in North America, and University of Iowa gastroenterologist Dr. Joel Weinstock thinks he may have an idea why.

Modern food-processing techniques and attention to hygiene have dramatically reduced the incidence of parasitic worms in humans. In the early 1900s, almost half of all children had worms, sometimes as long as 20 centimeters, living in their colon. The body's immune system kept the worms in check and essentially rendered them harmless. But as infection by parasites declined, the incidence of inflammatory bowel disease began to rise sharply. It seemed as if our immune system, when robbed of enemies to attack in the colon, began to attack the colon itself. So Weinstock hatched the idea of treating patients with a beverage made with pig whipworm eggs. In his most recent trial, 200 people suffering from ulcerative colitis or Crohn's drank the beverage twice a month. In the case of ulcerative colitis, half the patients went into remission from their abdominal symptoms, and 70 percent of the Crohn's patients showed remarkable improvement. A German company, BioCure is set to market the TSO (short for *Trichuris suis ova*) drink in

Europe. And who knows, in the future we may be giving our children whipworm-fortified foods to prevent IBD.

Downing an extract of okra, or "gumbo," as it is often known, sounds more appealing than slurping a whipworm brew. Andreas Hensel at the University of Dusseldorf found that this vegetable with the slimy flesh contains a mix of proteins and complex carbohydrates that can prevent *Helicobacter pylori*, the bacterium that can cause ulcers, from binding to the wall of the intestine. Plans are under way to add this okra extract to yogurt or muesli to prevent people from being infected with this nasty microbe.

Still, it may be a challenge to get people, especially children, to consume the newfangled functional foods. I had a hard time getting my youngest daughter to take vitamin pills. They taste "yucky," she complained. But then I sent for some "Vitaballs." These gumballs are packed with eleven essential vitamins, in the same dose as found in the "yucky" pills. You can even blow bubbles as you absorb the vitamins. It turned out to be a more pleasant experience than the one I had with my mother's foray into functional foods.

Eating Shellac

"And for my next trick . . . I think I'll eat some shellac." That line always gets a buzz out of the kids when I "perform" at an elementary school. I think they do like the color changes and the explosions, but they really come alive when they think you're going to eat something "gross." And the sticky secretion of the tiny Indian "lac" bug seems as gross as you can get.

Indeed, shellac is the resinous secretion produced by the female of the *Laccifer lacca* species. This little insect spends its whole life attached to a tree, sucking its juices, and converting them into the familiar sticky substance that has long been used to provide a glossy protective coating on wood. It takes a colony of about 150,000 insects to produce a pound of the resin, which has a variety of other applications ranging from stiffening hats to making buttons. The first hair sprays had shellac as their main ingredient, and the first phonograph records were made of this material. Shellac is soft and flows when heated, but becomes rigid at room temperature.

So you can see why, after hearing the apparently distasteful origins of shellac, a young audience would be excited by the prospect of my eating an "insect discharge." I must admit, though, they seem a tad disappointed when, instead of dipping into a can of varnish, I plop a shiny piece of gum into my mouth! The explanation that the gum is coated with shellac usually elicits the expected "yucks." In the food industry, shellac is often referred to as "confectioner's glaze," and can be used to give a protective, glossy coating to gum, candies, jelly beans, and ice cream cones. Since shellac is insoluble in water, it prevents the product from drying out by forming a layer impermeable to moisture. Citrus fruits and avocados are sometimes treated with shellac for this very reason.

Like any food additive, shellac is subject to rigorous safety regulations. Animal tests have shown no adverse reactions, and it has a long record of safe use in humans. This is not surprising, since the components of shellac—mostly a mixture of organic acids and esters—are found in a wide range of foods. What *is* surprising is people's—particularly children's—aversion to eating anything that derives from insects, with the possible exception of the sticky regurgitation of the honeybee.

Of course, just because honey happens to be purified bee barf doesn't mean we shouldn't eat it. Just as there is no reason to give up candies or gum because they may be coated with a bug secretion. But there certainly are reasons to limit intake of such sweets. For one, bacteria in the mouth can convert sugar to acids that cause tooth decay. Enter the "sugarless" gums and candies! Actually, they aren't really sugarless. They contain either sorbitol or xylitol, which are basically also sugars. Xylitol is particularly interesting. Its source may not be quite as unusual as that of shellac; it comes from corncobs, peanut shells, or birch bark.

The appeal of xylitol is not that it provides no calories. Actually, it does provide calories, although not as many as sugar. That's because xylitol is very slowly absorbed through the intestinal wall, and much of it is excreted. This is of real benefit to diabetics, since it means that the rise in blood glucose and insulin response associated with sugar is significantly reduced. But perhaps the most seductive feature of xylitol is that the *Streptococcus mutans* bacteria in our mouth cannot convert it to acids, and therefore, while it sweetens almost like sugar, it does not contribute to the formation of cavities. In fact, not only does xylitol not promote tooth decay, it actually reduces its occurrence. A recent study examined the effects of chewing gum sweetened with xylitol. Almost 300 eight and nine year olds were assigned to either a "no gum" control group, or a xylitol group where one stick of gum was chewed for five minutes three times a day on each school day for two years. It turned out that the gum chewers had significantly lower progression of tooth decay and less plaque on their teeth.

There is still more about the potential benefits of xylitol. One of the most common medical problems in young children is a bacterial or viral infection of the middle ear known as acute otitis media. It usually develops as a complication of the common cold when viruses reach the middle ear through the Eustachian tube from the throat. The viral ear infection is often followed by a bacterial infection. There is pain, high temperature, and the risk of complications, including infection of surrounding bone, hearing loss, or even the inflammation of the covering of the brain, known as meningitis.

The usual treatment is with amoxicillin, which works well. But obviously, it would be better to prevent the infection in the first place and cut down on the use of antibiotics. This may be possible by doing something as simple as chewing gum! Not any gum, mind you, but gum sweetened with xylitol. At least two major studies have shown that chewing xylitol-sweetened gum can reduce chronic ear infections. In preschoolers, ear infections were reduced by a whopping 50 percent!

Apparently, xylitol prevents bacteria from attaching to the back of the mouth from where they can later migrate to the ear and cause infection. Xylitol also inhibits the growth of bacteria, particularly *Pneumococci,* which is the species often responsible for acute otitis media. In the studies, about 10 grams of xylitol a day did the job, equivalent to about three sticks of gum. Perhaps worth a try for children who spend the winter coming down with ear infections and practically live on antibiotics. And there is a bonus. Chewing xylitol-sweetened gum is really cool. The compound has a large negative heat of solution, which means that it produces a refreshing cool taste in the mouth. Any potential harm with xylitol? In rare cases it can cause loose bowel movements. That's it. So how about some xylitol-sweetened gum? And don't let the fact that it may have a shellac coating bug you.

NASTY MICROBES

Do you know if your waiter sings in the bathroom? Or if the lady who whipped up the icing on your cake wore false fingernails? When was the last time you microwaved your dishcloth, or put your cutting board in the dishwasher? Is your orange juice pasteurized? Do you know if the shallots you ate in the restaurant were properly washed? Was the chicken thoroughly cooked? Let me tell you, these are more important questions than whether you eat fresh or farmed salmon, whether there is acrylamide in your French fries, whether genetically modified foods should be labeled, or whether fruits and vegetables harbor traces of pesticides. Why? Because we are not talking about theoretical risks; we are talking about real ones. The kind that can land you in a hospital, and potentially even destroy your life. We are talking about what microbes can do.

They rule the world, you know. We may think that we are in charge, but we're not. Microbes are. Without them, we would have no oxygen to breathe, no beer or wine to drink, no cheese or yogurt to eat. They help us digest our food, decompose our garbage, bake our bread, and synthesize our medications. And they also can make us sick. Very sick.

It often starts with a little tingling in the fingertips and toes. Then the sensation progresses up the arms and legs. Weakness sets in. Next comes paralysis and the inability to move anything except the eyelids. With luck, symptoms resolve within a couple of months. But a small percentage of patients die, and about 20 percent suffer permanent nerve damage. This is Guillain-Barré Syndrome (GBS)! And it can be caused by something as seemingly trivial as eating undercooked chicken.

Most cases of GBS occur after a bout with a bacterium or a virus. Somehow, the infection kicks the body's immune system into overdrive, causing it to attack the protective sheathing (myelin) around the nerves that connect the brain and spinal cord to the rest of the body. Often the culprit is *Campylobacter jejuni*, the bacterium that probably causes more food-borne infections in North America than any other. Most infections are associated with eating undercooked poultry or unpasteurized dairy products and result in nausea, vomiting, fever, diarrhea, and abdominal pain—the usual unpleasant symptoms of gastro-enteritis. But in roughly one in a thousand cases, *Campylo-bacteriosis* progresses to life-altering Guillain-Barré Syndrome. And that's not all. Sometimes, just like *Salmonella* infections, *Campylobacter* can cause "reactive arthritis," which may or may not resolve. Fortunately, thorough cooking readily destroys *Campylobacter*.

But how can we be sure that in a restaurant the poultry is thoroughly cooked, or that the cutting board used for poultry was adequately cleaned before slicing vegetables? There is not much we can do except rely on government inspectors to assure our safety. At home, though, we can do something. Statistics show that about a quarter of consumers do not clean their cutting boards after using them for chicken. Whether wood or plastic, they should either be washed in the dishwasher or cleaned with a dilute bleach solution. Washcloths are another problem. Microwaving a dry cloth for thirty seconds, or a wet one for three minutes, kills all bacteria.

Bacterial infections can sometimes come about in the most unusual fashion. Who would ever think that having breakfast with Mickey, Donald, and Pluto at Walt Disney World would send dozens of people to their physician with *Salmonella* infection? How could this happen at a theme park that prides itself on being squeaky clean? Well, it seems the plant where the unpasteurized orange juice served at the "character breakfast" was squeezed was not so squeaky clean. After the *Salmonella* outbreak was traced to the orange juice, inspection of the plant revealed cracks and holes in the ceilings and walls. And the toads that were found cavorting outside those walls turned out to be contaminated with *Salmonella*. They were likely responsible for the guests' misery. Walt Disney World restaurants now serve only pasteurized orange juice.

Even producers who use only pasteurized products can run into trouble because total control over suppliers is virtually impossible. One of the largest outbreaks of *Salmonella* poisonings in North America was caused by ice cream made from pasteurized ingredients. It turned out that the trucking company that delivered the pasteurized ice-cream mix also hauled liquid unpasteurized eggs. Of course, the trucks were supposed to be completely sanitized between loads, but to save costs, it wasn't always done. *Salmonella* from the eggs contaminated the mix and made thousands of people ill.

By contrast, the green onions served in a Mexican restaurant in Pennsylvania made only a few hundred people ill. This time, the contaminant was the hepatitis A virus. We can only guess at how the virus came to infect the onions, but given that it is usually transmitted via the fecal-oral route, there are some disturbing possibilities. Perhaps the irrigating water used on the Mexican farm where the onions originated was contaminated with manure, or maybe one of the farm workers handled the produce with dirty hands, or maybe a waiter didn't wash his hands properly after answering nature's call. It takes soap and effort to wash hands properly. The whole process should take as long as it takes to sing a couple of choruses of "Happy Birthday!"

And what about the fingernails? An outbreak of gastroenteritis that affected over 200 people in Georgia was traced to a colony of Norwalk-like virus growing under the false fingernails of an employee who had prepared icing for a cake. It seems even scrupulous washing doesn't remove all the microbes under these appendages. Obviously, we can't control every aspect of our food supply. But at home, at least, we can make sure that we keep our counters, cutting boards, and dishrags clean, that we thoroughly wash our vegetables and cook our chicken. And a couple of rousing choruses of "Happy Birthday," while washing hands can help to ensure more happy birthdays in the future.

The French Paradox

Wine producers undoubtedly raised a glass to toast the discovery! They were elated when Harvard researchers Konrad Howitz and David Sinclair showed that a compound found in wine had a possible life-extending effect. Granted, they only demonstrated the effect in yeast cells, but still, an increase in life expectancy by some 70 percent was pretty dramatic. There was more excitement when Marc Tatar at Brown University blended the substance into corn mush, fed it to his fruit flies, and found they lived 30 percent longer than untreated flies. And what was this life-extending dietary supplement? Resveratrol, a compound the scientific community had long been interested in, owing to indications that it was responsible for the reputed health benefits of red wine.

Howitz and Sinclair did not set out to investigate wine. They were interested in studying the aging process, for which yeast serves as a good model since yeasts with different longevity have been identified. Why did some yeasts live longer than others? The Harvard scientists traced the effect to a specific gene, named sir2 ("silent information regulator"), responsible for the production of an enzyme, appropriately christened "sirtuin," that had the ability to repair damaged DNA. And what determined the activity of this gene? Nutritional status, for one! When yeast cells were starved of nutrients, they produced more sirtuin and lived longer. This meshed with the long-standing observation that animals on calorie-restricted diets have a longer life expectancy.

Humans also produce a version of sirtuin, but most people rebel at the idea of upping their production of the enzyme if it means cutting their caloric intake to the verge of starvation. What's the point of living longer, they muse, if you are constantly so hungry that you wish you were dead? That's why Howitz and Sinclair decided to explore the possibility of increasing sirtuin levels by other means. So they devised a laboratory test that could measure the cellular production of sirtuin, and proceeded to bathe cells in various chemical solutions to determine the response. Lo and behold, the compound that led to the most sirtuin production was resveratrol, the very substance that had already reached a near-mythical status because of its presence in red wine. Had the life-enhancing potential of red wine finally been placed on a firm scientific footing? Not exactly.

You can't mention the connection between wine and health without bringing up the "French Paradox." How is it that the French—who smoke excessively and eat an abundance of high-fat cheeses, sugary pastries, and foie gras—have a lower heart attack rate than North Americans? Way back in 1819, pioneer cardiologist Samuel Black noted that angina was far more frequent in Ireland than in France, and attributed this to "the French habits and modes of living, coinciding with the benignity of their climate and the peculiar character of their moral affections." Given that French morals are not likely to offer protection against heart disease, we are left with "modes of living." And here, the most alluring possibility, of course championed by the huge French wine industry, is the consumption of red wine.

Although scientists generally are reticent to discuss the benefits of alcohol in public, there is little doubt that four to seven alcoholic drinks a week significantly cut the risk of a heart attack. Not only that, but this amount of alcohol can also improve sugar control for diabetics, reduce the risk of stroke and prostate cancer, and even improve cognition. So what's the worry? That the body's response to alcohol is in the form of a "J curve," meaning that a little drink is more healthful than abstaining or drinking to excess. Not much more than a single drink a day is linked to breast and oral cancers, and even the smallest amount of alcohol during the first trimester of pregnancy can lead to a smaller head circumference, a crude measure of brain size. Then there are the car wrecks, liver problems, and social tragedies linked to excessive consumption. So it comes as no surprise then that researchers are interested in determining just what components of alcoholic beverages are responsible for health benefits.

Resveratrol certainly would seem to be a candidate since it reduces blood clotting and, at least in laboratory studies, prevents cholesterol from being oxidized to a more dangerous form. But the Health Professionals Follow-Up Study at Harvard, which monitored over 51,000 subjects for twelve years, revealed that beer and liquor, three times a week, were more strongly associated with protection against heart attacks than red wine. It seems likely that health benefits are due both to alcohol and various other compounds, including resveratrol, found in beverages.

All the talk about red wine and its supposed wonder component has resulted in resveratrol pills appearing in health food stores. Unfortunately, though, resveratrol is an unstable compound when not in wine, so these supplements have questionable value. One product, Longevinex, claims to be packed in airtight capsules under nitrogen and has been shown to produce antioxidant effects on human cells in culture, but as of yet, it has not even been tested in mice.

In truth, the red wine evidence for the French Paradox is not strong. Maybe the answer to the mystery lies in a greater consumption of fruits and vegetables by the French or, more likely, in a lower calorie intake. In spite of their reputation for gastronomy, the French just eat less than we do. Or there may be no French Paradox at all! Some contend that French record keeping is different, and some cases described as "sudden death" are not attributed to heart disease. But here is something that has been shown: dipping apples in a resveratrol solution can significantly increase their shelf life. And you know the story about an apple a day. Here then is a way that a wine component can really improve our health. The rest of the stuff about the links between wine and health is pretty confusing . . . enough to drive a man to drink.

Jittery Goats and Coffee Beans

Do we blame the goat, or do we praise him? That is the question. I'm talking about the jittery goat, the one that, according to legend, belonged to Kaldi, the Yemeni goat herder. One day, some 1,200 years ago, Kaldi found his goat in a highly agitated state, darting back and forth, keeping everyone awake with frantic bleating. What could have caused this strange behavior? Kaldi wondered.

The inquisitive shepherd decided to follow the animal, and he soon found the root of the problem. Well, actually, the root wasn't the problem; the trouble was the bush that grew from the root. It seems the goat had become bewitched after feasting on the bush's strange, violet-colored berries. This was too much for Kaldi to handle, so he sought help from his religious leader, the Imam. Apparently, the sage had a scientific spirit, and made a brew of the berries. When he sampled the concoction, his heart began to race, and he suddenly felt very alert. The Imam had discovered the effects of caffeine! He named the beverage "Kahveh," meaning "invigorating," and we have been partaking of various versions of it ever since.

Today caffeine is the most widely used drug in the world, as about 80 percent of adults in Western societies consume it in coffee, tea, or soft drinks. Of course, when we drink coffee, we don't consume only caffeine. There are hundreds of other compounds found in the bean, and more form as a result of the roasting process. It's a good bet that at least some of these have an effect on our health. But what kind of effect?

First, let's raise a somewhat disturbing point. If coffee were a synthetic substance, it would probably not be allowed on the market! That's because it contains at least nineteen compounds that have been shown to cause cancer in test animals. Indeed, the natural carcinogens we ingest in coffee far outweigh the synthetic pesticide residues in our diet that frighten people so much. Yet we have no evidence at all that drinking coffee can cause cancer. An epidemiological study has even shown that coffee drinkers have a reduced risk of pancreatic cancer. Other studies have linked coffee with protection from liver and colon cancer. Why are we unaffected by the carcinogens in coffee? It is because they are not present in amounts anywhere near those that can trigger cancer in animals, and because coffee also contains a variety of antioxidants with anti-cancer effects.

So, no worry about cancer. What about heart disease? Well, two compounds in coffee, cafestol and kahweol, found in the oil droplets released from coffee beans by the brewing process, can stimulate the liver to make cholesterol. The problem, though, only arises with Scandinavian, Turkish, Greek, or "French press" coffees because the liberated oils stay in the coffee. Filter paper retains the oil droplets, which explains why filtered coffee does not elevate cholesterol. But even unfiltered coffee only becomes a problem at high doses. It takes about five cups a day to raise blood cholesterol by 10 mg/dL, the smallest change that can be reliably measured. Of course, there are people who drink that much, or more. Finnish men and women who regularly drink seven to nine cups of boiled coffee a day do have higher cholesterol levels. Espresso also contains cafestol and kahweol, but is consumed in such small volumes that the effect is insignificant.

Perhaps the most meaningful study to explore the link between coffee and heart disease was undertaken by the Harvard School of Public Health. No association was found between coffee intake and heart disease or stroke in over 45,000 health professionals who were followed for several years, even when more than four cups a day were consumed. This is comforting in light of a Greek study that showed that just one cup of coffee can temporarily make blood vessels more rigid. An interesting observation, but it's apparently of no consequence.

In spite of some scares, no significant scientific evidence links coffee consumption to high blood pressure, osteoporosis, birth defects, or fibrocystic breast disease. More than three cups a day, though, may increase the risk of rheumatoid arthritis. There is no question that caffeine can increase urinary frequency, and men with prostate problems should take this into account. And, of course, nobody contests that caffeine is a stimulant. In fact, the Canadian army has shown that soldiers who chew caffeine-laced gum are more vigilant at night and have improved shooting accuracy. This sits well with the military, because the motto of modern warfare is that if you own the night, you win the war!

Believe it or not, coffee may even help in the wars against both Parkinson's disease and type II diabetes. Several studies have shown a decreased risk of Parkinson's with increased coffee consumption. This neurological disease is caused by a deficiency of a brain chemical called dopamine, possibly brought about by the overactivity of adenosine, another neurotransmitter. Adenosine is known to decrease dopamine levels. Caffeine inhibits the activity of adenosine, so a connection to Parkinson's disease is plausible. Since adenosine can lull people to sleep, inhibiting its activity also accounts for caffeine's stimulant effect. As far as type II diabetes goes, surveys have shown that four or five cups of coffee a day can reduce the risk by some 30 percent. Chlorogenic acid in coffee seems to keep sugar from being absorbed from the gut into the bloodstream.

Basically, then, there are no grounds to the allegations that coffee in reasonable amounts is harmful. It may even be helpful. After all, it seems that coffee is the main source of antioxidants in the North American diet! University of Scranton chemistry professor Joe Vinson determined the antioxidant content of more than 100 foods and beverages and found that, when frequency of consumption was taken into account, coffee provided the most antioxidants. A single cup (240 mL, or 8 ounces) contains about a gram of flavonoids, which are established antioxidants. This revelation, though, should not be taken as encouragement to drink more coffee. We don't need the extra jitters. And if we get our antioxidants from fruits and vegetables, we get a good dose of vitamins, minerals, and fiber to boot. But it does seem that we don't have to worry about drinking moderate amounts of coffee, except perhaps for the spent grounds. What do we do with them? Apparently, they are excellent for removing the smell of elephant urine. And you never know when that bit of information may come in handy.

To Label or Not to Label, that Is the Question!

I place "consumer information" on a pretty high pedestal. In fact, these days, I spend much of my time trying to provide reliable scientific information to a public often confused by the deluge of apparently contradictory data. So it should come as no surprise that I'm a strong advocate of product labels that provide us with useful information. But, at the same time, I'm also wary of the ease with which inappropriate labeling can mislead consumers.

Let me begin with an interesting and disturbing case in point. A few years ago, a television "infomercial" touted the benefits of "Rio Hair Naturalizer," a product aimed mostly at black women wishing to relax their curls. The guest "expert" on the show spoke of the horrors caused by other hair straighteners that relied on "harsh chemicals," and then pointed to the product's label with its "all natural" and "chemical free" claims. The host followed up on this by leading the audience in a chant of "What do we want to be? Chemical free!" Obviously, the product was not chemical free; nothing but a vacuum fits that description. It was, however, "natural." The Rio hair relaxer was formulated with cupric chloride, a naturally occurring mineral. It relaxes curls, but also causes hair loss, green discoloration, blisters, and scalp burns. Over 50,000 women filed court complaints before the FDA put a stop to the nonsense by arguing that the label falsely claimed the product was chemical free.

Far less serious than this, but also misleading, are products that proclaim themselves to be "cholesterol free." Technically, the statement may be correct, since cholesterol is found only in animal products, and these foods do not contain any ingredients derived from animal sources. But there are two problems here. First, the insinuation that other similar foods may not be cholesterol free, and second, that being cholesterol free offers a significant health benefit. For example, a vegetable oil that screams "no cholesterol" on its label suggests that other such oils do contain the substance. A popular cookie may declare it contains no cholesterol while it supplies huge amounts of fats that constitute a greater risk for heart disease than dietary cholesterol does. If we are interested in reducing heart disease risk, we should push for labels that tell us not only the fat content per serving, but also how this is distributed. It would be great to know the "trans fatty acid" content as well as the ratio of omega-3 to omega-6 fats. A high ratio here may be protective against a host of diseases. Now *that* would be useful information.

And it is useful information that we hope to glean from a label, isn't it? That brings us to the thorny issue of labeling genetically modified foods. Suppose a label states "contains no GMOs." What message does that send? Does it not suggest that there is a reason to avoid GMOs? But genetically modified foods on the market have been approved by stringent regulatory agencies both in Canada and the US. They have been assessed for their effect on human and animal health as well as environmental safety. Not a single case of human disease has ever been attributed to any such food. But in spite of this, if consumers do want to see such labels, then the information has to be verifiable. This presents a problem: there may not be anything to verify, since there is no chemical difference between soy oil made from genetically modified soybeans and oil made from non-GMO beans. We may also see dishonest producers jumping on the bandwagon and labeling everything in sight as "GMO free," including products for which genetic modification is not an issue. Do we need to see apples or oranges labeled as GMO free when no apples or oranges are genetically modified?

What if we go in the other direction and focus on labeling foods that *are* sourced from genetically modified plants? Clearly, a bag of soybeans, canola, or corn grown with the aid of this technology can be so labeled. But what about a ready-to-eat meal that lists cornstarch as an ingredient? The starch may have come from corn with a gene inserted to produce the insecticidal Bt toxin, but the starch cannot be distinguished from starch that comes from non-GM corn. Would it be labeled? What about meat from animals raised on genetically modified feed? Or eggs from chickens similarly raised? How about a tomato modified with a gene from another variety of tomato? Or cheese produced using the enzyme chymosin that was made by recombinant DNA techniques, but is identical to chymosin found in the stomachs of animals? If regulations that require foods containing GM components to be labeled are introduced, what will be the maximum amount allowed to be present without such a label? 1 percent? 5 percent? 0 percent? How can this possibly be enforced? How will GM crops be kept separate? How much will this cost?

There does seem to be a reasonable way out of this conundrum. Label foods not according to the process by which they were produced, but according to the contents of the final product. If a genetically modified food is nutritionally or compositionally different from its traditional counterpart, it should be labeled as such. If there is no difference, then what is the purpose of labeling? This actually is the current point of view of the Canadian government and its scientific advisors.

So how do we respond to those consumers who say they have the right to know what they eat, even if there are no safety concerns? Fine, but why focus only on GM foods, then? What about asking for declarations about the number of insect parts or rat droppings allowed per serving (there are regulations about these), or the specific pesticides or fertilizers used, or toxins introduced by traditional crossbreeding, or whether the food was grown hydroponically? What about labeling lima beans as a source of natural cyanide? Why not put a warning on alfalfa sprouts about the risk of *E. coli* 0157:H7 poisoning? Shouldn't organic foods produced from crops sprayed with Bt bacteria be labeled? These bacteria release the same toxin as crops that have the Bt gene inserted. Don't consumers have the right to be informed about these things? Obviously, the labeling issue is not a simple one, and there are diverse views— though not of equal validity. Dr. Andrew Weil, whose views on "natural healing" have turned him into a veritable industry, suggests "not to buy products whose labels list more chemicals than recognizable ingredients." I wonder what he thinks "recognizable" ingredients are made of?

SMOKED MEAT

I'll admit it: I like smoked meat. And nobody needs to tell me that it is not "good for me." Still, I don't think that every morsel of food that slides down the esophagus has to be evaluated in terms of its nutritional value. I assure you that it is possible to indulge in this delicacy—not every day, mind you—

and still maintain a healthy diet. Just as it is possible to totally shun smoked meat and have a disastrous dietary regimen. True, smoked meat lovers would lose a nutritional debate to the bean sprout and brown rice warriors. But can the delight of biting into a well-stacked smoked meat sandwich be matched by slurping miso soup or chomping on tofu burgers? Not as far as I'm concerned.

Montreal is the smoked meat capital of the world. Period. I would venture to say, though, that most natives have no idea about how this famous product is made. It all starts with beef bellies from Alberta, which, in local lingo, we call briskets. The process of converting these to smoked meat begins by treating the briskets with "Chile saltpeter." And here a little history lesson is appropriate.

Perhaps the oldest of all food preservation techniques is "salting." Our ancestors discovered that treating meat liberally with salt slowed down the putrefaction process. Salt serves as a "dehydrating" agent, sucking water out of bacteria, and destroying them. But over the years it became apparent that some forms of salt resulted in a better product in terms of keeping qualities, color, and taste!

The reason was a natural contaminant of sodium chloride, namely sodium nitrate, or "saltpeter." Today we understand how it works. Microbes in the meat convert nitrate to nitrite. This species is a very effective antibacterial agent, especially against the potentially deadly *Clostridium botulinum* bacterium. It also reacts with myoglobin, a compound found in muscle tissue, to produce the appealing pink color of nitrosyl myoglobin. And, last but not least, sodium nitrite adds a characteristic "cured" flavor to the meat. Unfortunately, it also adds a health concern. Nitrites can lead to the formation of nitrosamines, which in animals have been shown to be carcinogenic. But more about that later.

The nitrate, which eventually yields nitrite, is dissolved in water and is injected into the meat by means of a specialized machine. Then comes the critical step in terms of flavor. The surface of the meat is rubbed with a blend of "secret" spices. There is salt in the mix, of course. In the old days it used to come in large grains called "corns," hence the expression "corned beef." Coriander, black pepper, chili powder, and bay leaves also add their flavor. Then there is freshly ground garlic! Not only is it an essential contributor to flavor, but also the sulfur compounds it contains have been shown to reduce nitrosamine formation. Finally the meat is packed into barrels, then cured in a fridge for two weeks.

Now for the all-important smoking process—except that it isn't really "smoking." In the old days, meat used to be hung in a smokehouse, exposed to all the compounds generated by burning wood. This cooked the meat, added flavor, and also served to preserve the meat. Chemicals in smoke, such as formaldehyde, are highly toxic to bacteria. That, of course, is why formaldehyde is used to preserve tissue specimens in the laboratory. Unfortunately, wood smoke also contains a number of compounds that are known to be carcinogens. So how can we smoke meat without worrying about these substances? The truth is that today there isn't all that much worry about the wood smoke because "smoking" is commonly done in a gas-fired oven, where the meat cooks by convection and the only smoke it is exposed to is generated by the fat that drips down from the meat and burns. This smoking process is still not free of concerns, since the high temperatures generate "heterocyclic aromatic amines," which are carcinogenic. But there may be a way around this problem, too. Just wait! Some commercial "smoked meats" are not smoked at all, but are injected with smoke flavoring. The less said about these, the better.

After about four hours in the oven, and once the meat has reached an internal temperature of 185°C (365°F), it is removed and sprayed with cold water to stop the cooking process. At this point, it is either vacuum packed, or placed in a refrigerator. Prior to eating, the meat has to be steamed for about an hour and a half to restore the water that has been lost during the smoking process. Then it is ready to be cut. And that is a job that requires special training. In Montreal, a "smoked meat cutter" is a highly respected professional, trained to cut against the grain of the meat to produce perfect slices.

Those slices may be perfect visually, but not nutritionally. There's the nagging matter of those nitrosamines, which can disrupt our DNA molecules and initiate nasty processes—perhaps even cancer. But studies have shown that chemicals in tomato juice, such as coumaric or chlorogenic acids, can inhibit this reaction. Research has also shown that vitamin C prevents the reaction of nitrites with amines in the food, or indeed in our bodies, to form nitrosamines. Therefore, an appetizer of tomato juice is great, and orange juice would seem to be the best beverage to accompany a smoked meat sandwich. Purists will surely rebel, claiming that anything other than a black cherry drink is sacrilegious.

Now, what about those heterocyclic aromatics that are byproducts of the cooking process? These form in cooked meat in amounts proportional to the temperature and cooking time. But you know what? Tea contains polyphenols, which have been shown to reduce the mutagenicity of these heterocyclics. Similar compounds are also found in apples. So why not cap off the meal with tea and an apple?

I guess you've gathered by now that there is a moral in here somewhere. Individual foods should not be vilified or sanctified. It is the combination of substances that we put into our mouth that determines our nutritional status. Indeed, smoked meat may not be the most nutritious food. But the nutritional concerns associated with it can be greatly reduced if it is harmonized with other foods and beverages. Unfortunately, pickles and French fries are not the most harmonious accompaniments. Not scientifically speaking, anyway. And please, New Yorkers, spare me the mail about the wonders of your pastrami and corned beef. I've had both. I've been to the Second Avenue Deli. I've been to Carnegie. I've been to the Stage. They may pile the meat sky high, and it isn't bad, but it isn't "smoked meat."

NUTRIGENOMICS

Let me make a prediction about your future. I predict that soon you'll become interested in a subject that you probably know nothing about right now. Nutrigenomics! What is it? It is the study of the interplay between our genes and our diet. Nutrigenomics will eventually allow you to walk into a physician's office, have your genetic makeup analyzed, and walk out with a prescription. But that prescription may not come in the form of pills in a bottle. It may be just a list of specific foods to eat and to avoid.

Let's begin with a concrete example. About one in 15,000 children is born with an inability to process phenylalanine, a common amino acid found in the diet. The gene that normally gives the instructions for making phenylalanine hydroxylase, the enzyme needed to metabolize phenylalanine, does not function properly. The result is phenylketonurea (PKU), a condition characterized by an accumulation of phenylalanine in the brain, possibly leading to brain damage. Treatment is simple enough. Foods that are high in phenylalanine must be avoided. So here we have a clear example of a link between genetics and nutrition. A genetic aberration sets up the possibility of a disease, but the disease only appears when certain foods are part of the diet.

Luckily, PKU is a rare condition. But heart disease is not. Here, too, both genetics and diet play a role. When a physician takes a medical history and asks about heart disease in the family, he is really performing a primitive genetic test. If the answer is yes, he will likely recommend a more stringent cholesterol-lowering diet than he would in the absence of such a family history. Of course, we know that not everyone benefits from such a diet. There are numerous cases of people who eschew fat, disdain eggs, and exercise religiously, only to suffer a heart attack. Indeed, about half the people who have heart attacks have normal blood cholesterol levels. However, as we learn more about the actions of genes in our body, we will be able to identify those who are likely to benefit from specific dietary recommendations. Like those people who carry a variant of a gene that codes for a protein with the tongue-twisting name of Apolipoprotein E.

Cholesterol is not an evil molecule, although it is clearly associated with heart disease. Actually, every cell in our body requires it. Cholesterol is both part of the cell membrane, and also the raw material cells use to manufacture sex hormones, bile acids, and a variety of essential biochemicals. But we do not need to consume cholesterol in our diet; the liver is capable of making it. Since cholesterol is not soluble in blood, it has to be transported around by attaching to molecules called lipoproteins. One of these is Apolipoprotein E. Like any protein in the body, it is made when the gene that "codes" for it is turned on. Unfortunately, though, there are several different forms, or "alleles," of this gene. One variant gives rise to a form of Apo E that deposits excessive cholesterol in the coronary arteries, where it can cause the buildup of plaque and lead to hardening of the arteries.

Finland has one of the highest rates of heart disease in the world, and also has one of the highest frequencies of the variant Apo E gene in its population. Japan has a low rate of heart disease, and the variant Apo E is rarely found in Japanese. However, the presence of the gene is not enough to explain the disparity in heart disease rates! The Finns may have a greater genetic susceptibility, but it is their high-fat diet that converts the susceptibility to active disease. In other words, a Finn who knew that he had this genetic variant would do well to pay very close attention to diet. The Apo E variant may be there, but it is less likely to cause damage if there is less cholesterol available to deposit.

The natives of Papua New Guinea present an interesting example. Here, for some reason, the frequency of the variant Apo E gene is extremely high, and one would expect the incidence of heart disease to be high as well. But it isn't. At least it hasn't been—until recently. That's because the diet, perhaps aside from the occasional plump missionary, has been extremely low in fat. Now, with a Western lifestyle encroaching, heart attacks at a young age are becoming common. Genes may deal you a poor hand, but lifestyle can still be an ace in the hole.

Of course, many genes—not just one—play a role in heart disease. Virtually every newspaper picked up a report from the *New England Journal of Medicine* that described the case of an eighty-eight-year-old gentleman who, for at least fifteen years, ate twenty-five soft-boiled eggs every day and still maintained a normal blood cholesterol level! He had a genetic variation that prevented cholesterol absorption from the gut. Wouldn't you like to know if you had such a genetic gift?

Heart disease is not the only condition that links genes and diet. There is evidence that autism, and even schizophrenia, may have such connections as well. A flaw in the production of an intestinal enzyme may lead to milk proteins not being completely degraded into amino acids in the gut, and result in the absorption of amino acid aggregates, which may have adverse effects on the brain. Then there is cancer. In some cases, a gene that codes for an enzyme, known as cyclooxygenase-2, is overactive and can lead to malignancy. However, resveratrol in grapes and curcumin in turmeric have been shown to inhibit the activity of this gene. So in the future, once your doctor has looked at your genetic profile, you may be told to drink grape juice and eat curry to reduce your personal risk of colon cancer. And how will the doctor do the genetic test? Believe it or not, a palm-sized instrument capable of analyzing your DNA is around the corner. Nutrigenomics, here we come!

The Saga of Golden Rice

You really will see better if you eat carrots. But there's a catch. Carrot therapy only works if your vision problems are due to a deficiency in vitamin A. This is a rarity in North America, but sadly not in the developing world. An estimated 250,000 to 500,000 cases of childhood blindness are caused by a diet that is deficient in vitamin A, or in its precursor, beta-carotene, which the body can convert to the vitamin.

What does vitamin A have to do with vision? Retinol, as the vitamin is also known, is absorbed from the digestive tract and is chemically modified to become retinal in the body. The retinal then complexes with a protein in the eye that is known as opsin. When light hits the opsin-retinal complex, a chemical change ensues, unleashing a cascade of events that lead to the transmission of an impulse up the optic nerve. Given that vitamin A is found in meat and fish, a dietary deficiency in North America is unlikely. Even vegetarians are safe. Although vitamin A occurs only in animal products, our bodies can make it from beta-carotene, the orange-colored molecule found in carrots and numerous other vegetables. Rice, however, has virtually no beta-carotene, and as a consequence, vitamin A deficiency in rice-based societies, such as India, China, and Indonesia, is common. As a result, these populations experience widespread childhood blindness, and tragically more than half of those who lose their sight die within a year.

Various attempts have been made to supplement the diet with vitamin A. In Indonesia, for example, it has even been added to packets of the widely used flavor enhancer MSG. But the problem persists. That's why so much excitement was generated in 2000, when recombinant DNA technology made possible the insertion into rice of a gene that codes for the production of beta-carotene. This gene, taken from daffodils, allowed the newfangled rice to produce enough beta-carotene to actually color it yellow—hence the term "golden rice."

Proponents of genetically modified crops highlighted the development of golden rice as a breakthrough and suggested it would be a useful way to put a dent into the vitamin A deficiency problem. Opponents pointed out that the amount of beta-carotene—roughly one and a half micrograms per gram of rice—was too little to have any practical impact. They claimed that the whole golden rice issue was an industry ploy to push for wider acceptance of genetic modification. Researchers countered that the technology was new, and that improvements would surely be forthcoming. And they were right! A team at Syngenta Seeds in Britain has found that a gene taken from corn and inserted into rice is far more adept at churning out beta-carotene than the original one from daffodils. This second-generation golden rice contains almost twenty-five times as much beta-carotene as the original version. A typical daily serving of 200 grams could then provide the minimal vitamin A requirement. Studies still have to be carried out to examine how cooking affects the beta-carotene content, however, and the efficiency of absorbing the nutrient from rice remains to be investigated. And, even though this is a truly remote possibility, any potential harmful effects will have to be ruled out.

Rice enhanced with beta-carotene can do more than help with visual problems. Vitamin A deficiency can lead to abnormal bone development as well as a greater susceptibility to infections. Low blood levels of vitamin A have even been linked with an increased risk of cancer. And should you think that Syngenta is just an evil multinational, trying to capture the rice market in developing countries with overly optimistic promises, know this: the company has donated the rights for golden rice to the non-profit Humanitarian Rice Board, which will make it available to farmers for free. India and the Philippines have already approved trial plantings, despite objections from anti–genetic modification groups that claim golden rice is a pie-in-the-sky approach and will not solve the vitamin A deficiency problem. But scientists have never claimed it would. Golden rice is just one method of providing extra vitamin A. And we do have to be impressed by the fact that, in just five years, researchers have found a way to increase the beta-carotene content of golden rice twenty-fivefold! Imagine the breakthroughs that the next few years may bring.

These potential advances are not limited to rice. In India, there is hope that a genetically modified potato will help combat malnutrition. Much of the population is vegetarian, but pulses and legumes, the main sources of protein, are expensive and often in short supply. Potatoes can be grown easily, but they don't contain much protein. This can be remedied, however, through the addition of a gene isolated from a South American plant known as amaranth. The gene in question codes for the production of a protein rich in the essential amino acids lysine and methionine. Too little methionine in the diet is

known to affect brain development. Amaranth is commonly eaten in South America, so the transfer of a gene from it into potatoes does not present a health risk.

Admittedly, many people remain suspicious of this new technology in spite of its potential to address nutritional problems. They don't want their food genetically modified. Little do they realize that virtually everything we eat has been modified, although not necessarily through the use of recombinant DNA. Centuries of crossbreeding, as well as treatment of seeds with chemicals or radiation to induce mutations, have resulted in extensive genetic modification of plants. In most cases, thousands of genes with unknown function may be involved. People don't worry about this (and shouldn't), yet they become extremely concerned when a single specific gene with a known function is transferred. I guess this isn't too surprising, given that a recent survey showed that 43 percent of Americans believe that only genetically modified tomatoes contain genes. Maybe if they ate more carrots, modified to contain vitamin A, they would see this situation better.

PESTICIDE PROBLEMS

Pesticides are designed to kill. Of course, they are designed to kill the insects, the fungi, the rodents, and the weeds that compete for our food supply, that carry disease, or that tarnish our green space. But they can also kill people. And, unfortunately, that isn't a rare occurrence. The World Health Organization estimates that there are roughly three million cases of pesticide poisoning worldwide every year, and close to a quarter-million deaths! Astoundingly, in some parts of the developing world, pesticide poisoning causes more deaths than infectious disease. How? Certainly, people do die from a lack of proper protective equipment, or because they can't read the instructions about diluting the chemicals properly. But the real tragedy is that the main cause of death due to pesticides is suicide!

Believe it or not, about a million people in the world do away with themselves every year. More than three-quarters of these are in Third World countries, where life can be so miserable that the alternative seems more attractive. In Sri Lanka, suicide is the number-one cause of death among young people, and in China, more young women kill themselves than die from other causes. Pesticides are the weapons of choice. In rural Sri Lanka, pesticide poisoning is the main cause of death reported in hospitals. There are wards devoted to patients who have tried to kill themselves with organophosphates, one of the most toxic classes of pesticides. In 1974, when paraquat was introduced in Samoa, suicide rates went up, sharply. They dropped back down in 1982, when paraquat was taken off the market. In Amman,

Jordan, poisonings fell way off when parathion was banned. Obviously, if the use of the most toxic pesticides could be curtailed in these countries, many lives would be saved. Sadly, though, these chemicals are often completely unregulated. Some of the most toxic ones are readily available in stores, and will be sold to the illiterate farmer who has virtually no chance of using them properly. Pesticide companies, in some cases, pay their salespeople on commission, so it is in their interest to push product even when it may not be necessary. In Sri Lanka, pesticides are advertised on radio to the public, often painting an unrealistic picture of magical, risk-free crop protection. Some sort of joint effort by pesticide manufacturers and governments is needed to keep the most toxic pesticides out of developing countries.

In North America, our pesticide regulations are far more stringent, and farmers must be licensed to use these chemicals. That doesn't mean we don't have problems. In North Carolina, for example, roughly 100,000 migrant workers are employed on tobacco, vegetable, fruit, and Christmas-tree farms. Many of them live in dilapidated housing next to the agricultural fields, and their homes and bodies are contaminated with pesticides. Metabolites of organophosphates commonly show up in their urine. This is not surprising, given that access to showers and clean clothes after working in the fields is limited. Even though there may be no immediate effects of such exposure, sufficient studies have suggested ominous links—between pesticide use and neurological problems, developmental delays, Parkinson's disease, and cancer—to cause concern. What's the answer? Elimination of agricultural pesticides is simply not an option. But providing workers with safe housing, clean clothes, showers, and above all, pesticide safety training certainly is.

Of course, working in the fields of North Carolina is not the only way to be exposed to pesticides. Garden-supply stores sell a wide array of such products. They are all "registered," meaning that they have undergone extensive safety evaluation. Risks should therefore be minimal, if the products are properly used. That, though, is a big "if." An often-quoted study at Stanford University found a link between Parkinson's disease and domestic pesticide use. People with as few as thirty days of exposure to home insecticides were at significantly greater risk; garden insecticides were somewhat less risky. Because of the large variety of products available, the researchers were not able to zero in on any specific ingredients. Another study, this one at the University of California at Berkeley, compared pesticide exposures of children diagnosed with leukemia to a healthy control group matched for age and socioeconomic status. The families of children with leukemia were three times more likely to have used a professional exterminator. During pregnancy, exposure to any type of pesticide in the home coincided with twice as much risk. But—and an important "but"—there was no association between leukemia and pesticides used outside the house! Yet I have often seen activists who oppose "cosmetic" lawn-care chemicals use the leukemia argument to demonize this practice.

Pesticides cannot all be lumped together, in terms of their safety profile. There are tremendous differences between the various insecticides, which differ extensively from herbicides and fungicides. And one must always remember that associations cannot prove cause and effect. Physicians should realize this, one would think. Apparently, not all do. In a letter to a medical publication, a doctor chastised the federal government for allowing people to be exposed to dangerous substances on their lawns, and buttressed the argument with this example: "A boy was removed from a day care three years ago because his parents noticed the lawn was being treated with pesticides and the child began to suffer health problems and recurrent pneumonias. He developed acute lymphoblastic leukemia." The simple-minded message, of course, is that the spraying caused the leukemia—a gigantic, and inappropriate, leap of faith.

Great caution must be used with insecticides in the home, and I believe that their use during pregnancy should be totally avoided. But using insecticides inside a house presents a completely different scenario from occasionally spraying a lawn with fertilizer and weed killer. Different chemicals, different exposures, different risks. When contemplating the use of pesticides, always remember that, while there may be no completely safe substances, there are ways to use substances safely.

ORGANIC AGRICULTURE

There were piles of all sorts of tomatoes in the produce aisle of the supermarket. The ones that caught my attention sat neatly wrapped in plastic in groups of four. They weren't any better looking than the others, but their price was a stunning $5.80! What sort of tomatoes were these, to command a king's ransom? Well, they were "organic." Why did they warrant the investment? Because, as the label declared, "when you purchase organic produce you are taking part in the healing of our land, the purifying of rivers, lakes, and streams, and the protection of all forms of life from exposure to chemicals used in conventional farming." Surely only a callous chemist with a disregard for nature would purchase any other sort of tomato.

There is no doubt that the organic produce market is growing. Some buy organic because they believe such foods are healthier; others do so to help save the environment from those nasty agro-chemicals. These beliefs are certainly worth investigating. But what exactly does "organic" actually mean? Essentially, organic food must be produced without the use of synthetic pesticides, artificial fertilizers, antibiotics, or growth-promoting hormones. Genetically modified organisms are not allowed, and irradiation cannot be used to control bacteria. Sounds just like farming roughly 100 years ago. Back then, feeding the masses required the involvement of some 70 percent of the population in farming in some way. Yields were low, crop losses to insects, fungi, and weeds were high. That's why farmers welcomed the introduction of scientifically designed fertilizers and pesticides. That's why, today, 2 percent of the population can feed the other 98 percent.

Such advances have not come without a cost. Pesticides and nitrates from fertilizer enter ground water, posing potential environmental and health consequences. So people hark back to the "good old days," when food was untainted and people lived in blissful health. Of course, those "good old days" only exist in people's romanticized imaginations. Food-borne diseases were rampant, and fresh fruits and vegetables in winter were virtually unheard of. Nutrient deficiency diseases cut a wide swath through the population. Of course, not even the greatest advocates of organic agriculture suggest that we can realistically turn back the clock and provide food for the world's population using only organic methods. They claim a niche market that caters to people who are conscious of their environment and health.

So, do consumers who buy "organic" avoid pesticides? Hardly. Organic farmers are allowed to use a number of pesticides, as long as they come from a natural source. Pyrethrum, an extract of chrysanthemum flowers, has long been used to control insects. The Environmental Protection Agency in the US classifies it as a likely human carcinogen. There you go, then, a "carcinogen" used on organic produce! Does it matter? Of course not. Just because huge doses of a chemical, be it natural or synthetic, cause cancer in test animals does not mean that trace amounts in humans do the same. Furthermore, pyrethrum biodegrades quickly, and residues are trivial. But that is the case for most modern synthetic pesticides, as well! And how about rotenone? This compound was discovered in the 1800s in the extracts of the root of the derris plant. Primitive tribes had learned that, when spread over water, the ground root would paralyze fish, which then floated to the surface. Rotenone is highly toxic to humans and causes Parkinson's disease in rats. Organic farmers can use it to control aphids, thrips, and other insects on fruit. Residues probably pose little risk to humans, but synthetic pesticides with the same sort of toxicological profile have been vilified.

Organic farmers are also free to spray their crops with spores of the *Bacillus thuringiensis* (Bt) bacterium, which release an insecticidal protein. Yet organic agriculture opposes the use of crops that are genetically modified to produce the same protein. Isn't it curious that exposing the crop to the whole genome of the bacterium is perceived to be safe, whereas the production of one specific protein is looked at warily? The truth is that the protein is innocuous to humans, whether it comes from spores sprayed on an organic crop or from genetically modified crops. True, organic produce will have lower levels of pesticide residues, but the significance of this is highly debatable.

A far bigger concern than pesticide residues is bacterial contamination, especially by potentially lethal *E. coli* 0157:H7. The source is manure used as a fertilizer. Composted manure reduces the risk, but any time manure is used, as is common for organic produce, there is concern. That's why produce should be thoroughly washed, whether conventional or organic. Insect damage to crops not protected by pesticides often leads to an invasion by fungi. Some fungi, like fusarium, produce compounds that are highly toxic. In 2004, two varieties of organic cornmeal had to be withdrawn in Britain because of unacceptable levels of fumonisin, this natural toxin.

Are organic foods more nutritious? Maybe, but marginally. When they are not protected by pesticides, crops produce their own chemical weapons. Among these are various flavonoids, antioxidants that may contribute to human health. Organic pears and peaches are richer in these compounds, and organic tomatoes have more vitamin C and lycopene. But again, this has little practical relevance. When subjects consumed organic tomato purée every day for three weeks, their plasma levels of lycopene and vitamin C were no different from those seen in subjects who consumed conventional purée. Where organic agriculture comes to the fore is in its impact on the environment. Soil quality is better, fewer pollutants are produced, and less energy is consumed. But we are simply not going to feed 7 billion people organically.

Finally, do organic tomatoes taste better? I can't tell you. Instead of shelling out $5.80 for four tomatoes, I bought a bunch of regular tomatoes, some apples, and some oranges for the same total. And I think I got a lot more flavonoids and vitamins for my money.

Bringing Piggies to Market

I must admit, I had never heard of a "boar limo." Nor did I know about "Prosperm," "pit additives," or the risks of "plug pulling." But when you sit around a table with a bunch of pork producers, you learn quickly. And when you find out that the lady sitting next to you can castrate a boar in 1.5 seconds, you pay attention to the conversation. You learn how hard these farmers work, how daily life centers on feed costs, pork futures, worries about bacteria, concerns about smells, and insecurity about income. But you also find out how remarkably "scientific" pork production has become, and how extensively animal welfare and environmental concerns are addressed.

These days, pork production generally begins with artificial insemination of the sow. Farmers can purchase a variety of sperm to match their needs, but need to know exactly when a sow is in heat to maximize the chance of breeding. The best indication of heat is the so-called "immobilization response," whereby the sow's ears become erect (they "pop" in trade lingo) as she assumes a rigid position, ready to be mounted. But sows certainly did not evolve to have romantic relationships with artificial inseminating rods, so a boar is still needed. Or at least his smell is. Boars produce a mixture of compounds, of which androstenol and androstenone are the most dominant, to trigger mating behavior in the female. These compounds make for an unpleasant "boar-taint" in meat, the reason most boars are deprived of their testicles soon after birth.

Farmers can detect heat in a sow by guiding a boar (one with testicles intact) down the aisle between the stalls that house the females as they watch for a response. This is the process made more efficient with a "boar limo," the remote-controlled, motorized cart into which the boar is loaded for his romantic journey. As the farmer watches from behind the female, he maneuvers the limo to allow full "snout to snout contact," taking advantage of the four key factors needed for sow stimulation: sight, smell, sound, and saliva.

Producers who don't want to mess around with a boar to detect estrus can avail themselves of "Boar Mate," an aerosol spray that contains androstenol and androstenone. If a sow is in heat, she will assume the mating position when Boar Mate is sprayed in front of her nose, and the insemination rod then goes into action. The success of the insemination, as well as the size of the litter produced, depend on many factors, including

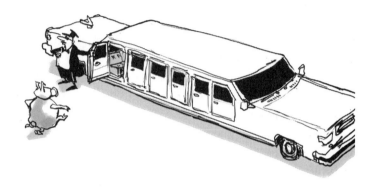

quality and quantity of sperm. This is where products such as "Prosperm" come in. This fertility enhancer can be fed to boars as a dietary supplement. Its marketers claim a significant increase in effective matings and in piglets per litter. What is this magic formula? Docosahexaenoic acid (DHA), a fatty acid that is an essential component of sperm cells, combined with vitamin E and selenium to protect the DHA from oxidative damage. DHA is readily isolated from tuna. Needless to say, some unethical entrepreneurs already promote such products to men who are concerned about their prowess, and suggest that the "Tuna Love Pill" can perform miracles. They cite a 15 percent increase in boar sperm count, and a testicle size increase of some 20 percent. As far as I can tell, the scientific literature provides no data on whether tuna eaters are better lovers.

Feeding pigs is not a simple business. Not because they are finicky eaters; in fact, pigs will eat most anything. Cost and nutrition are the issues here! Soy, corn, oats, barley, peas, lentils, and canola are the common feed components, but these differ in price and protein content, so they have to be judiciously selected and blended. Pigs, like humans, need amino acids to synthesize muscle tissue and enzymes. These come from breaking down dietary protein into its component amino acids. But if the amino acid ratios are not exactly what the pigs need, some of the excess will be excreted in the feces. This can be a problem, because amino acids can be a source of both nitrates and ammonia, which are environmental concerns. An effective approach is to add amino acids, such as lysine or methionine, to the feed in appropriate amounts, but the viability of this depends on relative costs.

One of the greatest concerns of a pig facility is the copious amount of manure the animals produce—up to 8 pounds a day, each. The stuff falls through the slats in the floor of the piggery and collects in tanks underneath. Periodically, the farmer pulls a plug, allowing the liquid manure to flow into outdoor lagoons. This is a potentially dangerous process, because anaerobic fermentation produces toxic hydrogen sulphide gas, which is liberated into the air when manure is agitated. The lagoons are designed by engineers and use various high-tech liners to ensure that there is no leakage. No manure is dumped into lakes or rivers, and groundwater around lagoons is constantly tested.

Pig manure makes for a highly effective, "organic" fertilizer, and farmers pump it into the soil to raise crops. Pig sludge can even be dried, mixed with wastepaper and sawdust, and burned for energy production. Waste management systems can be designed to capture the methane gas produced by decomposing manure, which in turn can be burned to produce electricity.

No doubt about it, though, the smell of manure is a huge problem. Trees planted around manure lagoons and "pit additives," such as certain enzymes and copper compounds that break down odiferous compounds, can help, as can the use of zeolites (a special form of volcanic rock) that can absorb smells. Odors are not only a problem for the neighbors; they are worrisome for the farmer as well. A buildup of ammonia in a barn is dangerous to the health of piglets, and there is concern about some of the "endotoxins" produced by bacteria that are housed by pigs. Inhalation of these vapors can cause severe respiratory problems. Antibiotics in the feed control bacteria, but raise the issue of developing resistance to these drugs. To prevent this, stringent laws are enforced, and animals with even a trace of antibiotic residue cannot be marketed. Farmers are subjected to crippling fines if they do not adhere to the regulations.

Even if the animals have been successfully raised, there is the matter of transportation on their final journey. A study in *The American Journal of Public Health,* with the tantalizing title of "Salmonella Excretion in Joyriding Pigs," concluded that stressed pigs release higher levels of salmonella. Bringing piggies to market is not a simple business.

CALCIUM AND WEIGHT LOSS

I normally just roll my eyes when I come across diet books festooned with the word *revolutionary.* That expression is usually the hallmark of some gimmicky regimen that fails to reduce weight but manages to increase the author's bank account. So it was with my usual skepticism that I began to look through *The Calcium Key: The Revolutionary Diet Discovery that Will Help You Lose Weight Faster.* In truth, the only reason I had any interest was because the author, Dr. Michael Zemel, is a professor in the Department of Nutrition and Medicine at the University of Tennessee, and has published numerous research papers in respected journals. He should know what he is talking about. And what he talks about is the connection between calcium intake and weight loss.

Dr. Zemel's interest in the calcium connection was aroused in the 1980s, when scientists began to explore the link between calcium and blood pressure. It became quite clear that people who had a high calcium intake had lower blood pressure. Unexpectedly, though, the researchers also noted that high calcium intake correlated with a lower body weight. Why should this be so? Zemel decided to find out. He had no trouble enlisting a few dozen subjects for his studies, for the simple reason that the subjects were mice. But these creatures do serve as good models for humans, as far as metabolism goes. Zemel and colleagues found that a diet low in calcium causes the kidneys to release calcitriol, a form of vitamin D. Calcitriol's role is to compensate for a low calcium intake by decreasing calcium excretion from the kidneys, increasing its absorption from the gastrointestinal tract, and stimulating its release from bones. This was no great surprise, since it was already well known that calcium is critical for heart and nerve function, and that the body has means of increasing blood levels when the diet isn't providing enough of the mineral. What did come as a shock was the discovery that calcitriol also increased calcium absorption into fat cells, where it triggered a cascade of chemical reactions. The end result was a buildup of fat in these cells.

There seems to be a reasonable rationale for this observation. When calcium intake is low, the body assumes that there is not much food coming in. So it takes emergency measures, and starts to store fat for use during the lean times that may be approaching. A nice theory, but could it be confirmed by evidence? Once again, Dr. Zemel reached for his mice. First, the mice, specially bred to be obese, were allowed to dine until they became portly. Then they were put on a low-calorie diet, with varying amounts of calcium. With no calcium, the mice managed to lose only 8 percent of their body fat. When they were given calcium carbonate supplements (à la "Tums"), the animals managed to shed 42 percent of their body fat. But the really exciting results came with the mice that had been given high amounts of low-fat dry milk. They led the pack with a stunning

69 percent loss in body fat! Mice, though, are not humans. Would people show the same effect? There was only one way to find out.

This time, the Zemel group enlisted thirty-two obese adults and planned a randomized, placebo-controlled trial. For twenty-four weeks, the subjects were put on a diet that contained 500 fewer calories than they were accustomed to eating, but with varying amounts of calcium. One group of subjects had at most one dairy serving a day, for a total intake of 400 to 500 milligrams of calcium, a second group had the same diet but also received an 800-milligram calcium supplement, for a total of 1,200 to 1,300 milligrams of calcium, and a third group was given three low-fat dairy servings, again for a total of 1,200 to 1,300 milligrams of calcium. The results were clear: the greatest weight loss, a pound a week, was seen with the high-calcium diet. Why calcium from a dairy source should be more effective than calcium supplements is not clear. Milk is chemically complex and contains numerous compounds that can apparently enhance the "calcium effect." Obviously, more research is needed in this area, and it will need to be supervised by scientists who do not have a financial stake in the calcium connection. In addition to profiting from his book, Dr. Zemel holds a patent on treating obesity with a high-calcium regimen.

Zemel's results are, however, corroborated by an epidemiological study carried out at Laval University in Quebec City. Here, researchers analyzed the nutrient intake of 235 men and the same number of women over seven years, and found that a low calcium intake was associated with obesity, particularly in women. Furthermore, they found that cholesterol levels were significantly lower in people who had a high calcium intake.

Now, then, what do we take away from all this? First, when we're dieting, we don't take away all dairy products, as so many people try to do. But we should take away the fat from the dairy products. Low-fat dairy has as much calcium as the high-fat version, so stick to it. A couple of glasses of skim milk a day will go a long way toward satisfying the daily calcium requirement. Remember, though, that calories count! Nobody is suggesting that adding low-fat dairy products to an ordinary diet is going to shed pounds. But incorporating such into a low-calorie diet just may enhance weight loss. And it will also benefit the bones, help control blood pressure, and even reduce the risk of colon cancer. Dr. Zemel's research may not exactly be "revolutionary," but it may represent a small victory in the battle against the bulge.

LEARNING FROM THE BUSHMEN OF THE KALAHARI

They're poor. They're uneducated. They have problems with alcohol and drugs. They certainly don't work in research labs. Yet the San Bushmen of the African Kalahari Desert may have set scientists on the track of finding an effective weapon in the battle against obesity. If you remember the surprise 1980 hit movie, *The Gods Must Be Crazy,* you'll be familiar with the Bushmen. If not, rent the video! These natives of the Kalahari are among the earliest known hunter-gatherer civilizations, but had no contact with Westerners until the 1920s. The film tells the story of a tribe living a peaceful life until a pilot carelessly drops a Coca-Cola bottle out of his airplane. The natives have never seen anything like this, and get into all sorts of conflicts about what to do with this magical gift. Finally one of the Bushmen decides that the bottle has caused so much discord that it should be returned to where it must have come from, the Gods. And so, as he embarks on his quest, a slew of zany misadventures begins.

Well, it seems that real life may have imitated art. The gift from the Gods, though, is not a bottle, but a plant. The San hunters have been using a type of cactus, *Hoodia gordonii,* for thousands of years to curb the appetite while on long forays in search of game. There wasn't much food to be had on these hunting trips, and the Bushmen discovered that snacking on the Hoodia cactus would make hunger pangs go away. Sometime in the early 1960s, rumors about the existence of this appetite-suppressing plant filtered down to scientists at the South African Centre for Scientific and Industrial Research (CSIR). Could there be something here to help deal with the rapidly expanding waistlines of Westerners who had become far too adept at hunting and gathering in the aisles of supermarkets?

It didn't take long to discover that rats given Hoodia extracts stopped eating. So maybe there really was something to those stories about Hoodia dispelling hunger pains! CSIR scientists began the task of isolating and identifying the "active" ingredient in Hoodia, finding the proper dosage, and determining its safety profile. It took some thirty years, but finally a compound, code-named P57AS3, was found to have appetite-suppressant properties. CSIR promptly patented it as a weight-loss substance, but did not have the means to produce significant amounts or to explore commercialization.

Enter Phytopharm, a British company with expertise in developing drugs from natural products. CSIR worked out a deal that granted Phytopharm a license to commercialize Hoodia. In turn, Phytopharm, realizing that it may not have all the resources needed to bring a project with such great promise to fruition, entered into an agreement with Pfizer, the giant pharmaceutical company. The stage was now set to see if Hoodia could really perform. As usual, rats were the first guinea pigs. Given Hoodia extract, they lost weight even on a highly palatable diet. Excitement mounted further with the first human trial. Twenty overweight subjects were confined to a metabolic ward for two weeks and given either powdered Hoodia P57 extract or a placebo. They were allowed to eat as much as they wanted as they sat around, read, or watched TV, just like many Westerners do. The results were startling! Subjects treated with Hoodia reduced their caloric intake by some 1,000 calories a day and lost about 2 kilos over the two weeks. Granted, this was only one short-term study, but it was encouraging, nevertheless.

The plot thickened further when researchers at Brown University Medical School injected the purified Hoodia extract into the brains of rats and discovered chemical changes in their hypothalamus. This is the part of the brain that maintains the body's "status quo" by controlling blood pressure, temperature, electrolyte balance, and weight. Injection of P57 resulted in an increase in the levels of adenosine triphosphate (ATP), the body's main energy storage molecule. The implication is that Hoodia tricks the brain into thinking that there has been an energy input from food, and a message is sent out to reduce intake. Surprisingly, in 2003, Pfizer decided to abandon the project, likely because the active ingredient, P57, turned out to be a complicated steroid derivative that was very difficult to synthesize on the scale needed to make standardized weight-loss pills.

The only viable approach now seemed to be incorporating Hoodia in its natural form into shakes, bars, or meal replacements. Phytopharm found a willing partner in Unilever, the multinational company that was already into the weight-loss game with its line of SlimFast products. Hoodia is now being grown on large, well-guarded plantations in South Africa to meet the expected demand, should further research bear out the early findings. The estimated market potential is over $3 billion a year! As is often the case, the hucksters have jumped on the bandwagon trying to cash in on Phytopharm's research.

Numerous Hoodia products already clamor for dieters' attention on the web. Analyses have shown that they contain essentially no active ingredient. "Bio-pirates" are also active in Africa, and attempt to steal Hoodia plants to feed Americans' appetite for anti-appetite substances.

And where does all this leave the San Bushmen? After all, it was their traditional knowledge that led to the discovery. For years they were shunted aside, but an agreement has now been reached that will allow the San to share in the profits, if indeed there are any. At this point, numerous stumbling blocks may still be encountered. But this hungry, poor, Third World tribe may have provided a partial solution to the problems of a developed world that is now eating itself sick. And that may not be all. Hoodia also has legendary aphrodisiac qualities. In the words of Petrus Vaalbooi, the bushman the San elected to look after their interests, "It's very good for men's problems. Once you've eaten this, you can really give your wife a good seeing-to." Hey, maybe those Gods weren't so crazy after all!

THE CHEESECAKE FACTORY

Let me tell you about Dr. Roy Walford and his anti-aging theory. But first, let me tell you what prompted this little discussion. My wife and I were walking around Coconut Grove, near Miami, when I spotted an outlet of The Cheesecake Factory. I had heard that this was a restaurant that must be tried, although I didn't quite remember why. It didn't take long to find out. We were ushered to a table for two and handed a menu almost as thick as a telephone book. This was the first clue that we were in for a unique adventure. Another clue sat at the table next to us, in the form of four diners who had obviously not wasted much energy over the years on any activity other than the

furious wielding of cutlery. This was clearly not a place for small appetites.

We decided to share a meal and ordered a salad, which I figured, judging by its $15 price tag, would be of significant size. The waitress seemed somewhat puzzled by the concept of two people sharing a meal, apparently not a common occurrence at The Cheesecake Factory. Hard to understand why, since the huge oval plate that arrived was piled with the annual yield of a small farm. But this was nothing in comparison to the feast that was laid before our neighbors by two waiters who had undoubtedly trained for their job through years of weight lifting. Those plates were heaped with fries, onion rings, ribs, burgers, and other assorted animal parts, to a height of at least 6 inches. Several cows, pigs, and chickens must have been involved in the process. I figured each dish amounted to at least 4,000 calories.

That was before the cheesecake. There were about twenty varieties on the menu, and our neighbors each had a piece. No sharing here! Needless to say, these were no ordinary slices of cheesecake. Just one could have fed half the Moldavian army for a month. I'd estimate that they contained at least 1,000 calories each. As I looked around the restaurant, I saw similar scenes being played out at other tables. Diners seemed to be racing to see who could burst out of their clothes first. I found The Cheesecake Factory pretty interesting, but I think Dr. Roy Walford would have had another description for the place. He would probably have called it a nightmare.

Walford was an absolutely fascinating character. His bald head and Fu Manchu mustache didn't conjure up the image of a professor of pathology, which is exactly what he was at UCLA medical school. His main research interest was aging, and how not to do it. While Dr. Walford never thought he had found the fountain of youth, he did think he had the answer to extending life expectancy by a couple of decades. And you don't have to take a mound of supplements to do it. All you have to do is forsake a few calories! Well, possibly more than a few: up to 30 percent of our average intake.

Walford based his theory on numerous animal experiments he carried out. He showed that rodents on calorie-restricted diets live longer and are less prone to cancer and kidney disease. Although certain cancers, such as those of the breast and the colon, have repeatedly been linked to a high-fat diet, it may be that total calorie intake, rather than fat, is the real culprit. Experiments have indeed shown that tumor formation in rodents can be prevented by a low-calorie diet.

One always has to be careful, however, in applying animal models to humans. But Walford cited the example of the inhabitants of the Japanese island of Okinawa. The natives there consume some 30 percent fewer calories than other Japanese, and have significantly reduced rates of cancer, diabetes, and heart disease. The island also has more centenarians than any other area of Japan.

Japan has the longest life expectancy in the world, and it is noteworthy that the typical diet adds up to only about 2,600 calories, the fewest of any industrialized nation. North Americans consume an average of 3,500 calories a day. There is certainly a difference in the fat content of the diet; 25 percent of the calories in the Japanese diet come from fat as compared with 45 percent in the American diet. But the real difference just may be the total calories consumed. Dr. Walford certainly thought so.

In fact, Dr. Walford believed in his theory so strongly that he made himself his own guinea pig. He consumed only about 1,500 to 2,000 calories per day, and sometimes even fasted on alternate days. He ate raw fruits, whole grains, and small amounts of animal protein. To ensure that he was only under-nourished, and not malnourished, Dr. Walford took vitamin and mineral supplements. He was going to show the world the benefits of his regimen! Unfortunately, amyotrophic lateral sclerosis (Lou Gehrig's disease) got in the way, and Walford died in 2004, at the age of seventy-nine.

Perhaps Walford's ideas were a bit extreme, but there is no doubt that many North Americans need to curb their caloric intake. Governments are looking at various ways to do this, and might even resort to legislation if necessary, but some members of the food industry are already mobilizing to stop the campaign of the "food nags." They want to be free to stuff people with chips, cookies, and sugared drinks to their heart's delight. On the other hand, lawyers are preparing to file lawsuits against big food companies, just as they did against the tobacco companies, for knowingly undermining the health of people. It will be an interesting era.

And how did our Cheesecake Factory experience turn out? The salad was excellent, and plenty for two. Then we shared a cheesecake, and couldn't finish the slice. I asked for a doggie bag. This again puzzled the waitress. I guess she had never before seen a customer who didn't polish off the mountain of food that was served. As we left, I noticed our neighbors were still sitting at their by then empty table. I think they couldn't get up.

A Bump for the Antioxidant Bandwagon

Science can drive you to despair. Just when you've become comfortable on the vitamin supplement bandwagon, convinced that a one-a-day multi is the way to go, you hit a bump on the road. A pretty large one! The bump in this case comes in the form of a headline, as featured on the cover of *The Lancet,* one of the prime medical research journals in the world: "The prospect that vitamin pills may not only do no good but also kill their consumers is a scary speculation given the vast quantities that are used in certain communities." Scary, indeed. What's going on?

The study that prompted the editorial comment on the cover page and stunned the scientific community appeared in the October 2, 2004, issue of *The Lancet*. Researchers led by Goran Bjelakovic of the University of Nis (Serbia and Montenegro) were interested in determining whether antioxidant supplements reduced the risk of gastrointestinal cancers and mortality. Both the lay press and the scientific publications have been singing the praises of antioxidants for years. Free radicals, those rogue species that roam through our body, damaging essential molecules like DNA, can be swallowed up by antioxidants, they claimed. At least that's what happened in the test tube. Furthermore, eating fruits and vegetables, which are loaded with the likes of

beta-carotene, vitamin C, vitamin E, polyphenols, and selenium—all recognized antioxidants—is clearly associated with a
reduced risk of cancer. So it certainly seemed reasonable to
assume that supplemental antioxidants should be beneficial in
preventing cancer, especially of the digestive tract, given that
this is the first part of the body they encounter. Numerous
researchers had explored this possibility, mounting studies that
compared outcomes between people taking vitamin supplements
and those taking placebos. Now Bjelakovic and colleagues
decided to pool these studies in a "meta-analysis," a "study of
studies." This is recognized to be an effective way to draw conclusions from trials that, by themselves, provide ambivalent
results.

They started by scouring the medical literature and various
databases for any study that had investigated the link between
gastrointestinal cancers and antioxidant supplements. There were
loads of these, but only fourteen trials fulfilled the scientific rigor
the researchers were after. In all, over 170,000 people were
involved in these trials, all of which were appropriately randomized and included placebo controls. Oral antioxidants were
used in all cases, although dosages varied, as did combinations.
Beta-carotene ranged from 15 to 50 milligrams, vitamin A from
1.5 to 15 milligrams, vitamin C from 120 to 2,000 milligrams,
vitamin E from 30 to 600 milligrams every day, or on alternate
days for one to twelve years. Selenium supplements (50 to 228
micrograms) were taken every day for two to four years. These
doses are not inordinately large, and are in the range of what
most people who take dietary supplements consume.

To everyone's surprise, the supplements offered no protection against esophageal, gastric, colorectal, pancreatic, or liver
cancer. The sole positive finding was that, in four of the trials,
selenium supplementation showed a significant beneficial effect.
But surprise changed to shock when, in seven high-quality trials

involving over 130,000 subjects, an increase in mortality was seen among supplement takers! Based on the data they had unearthed, the researchers calculated that for every million people taking combinations of antioxidants, about 9,000 premature deaths were to be expected. Given that millions of North Americans take such supplements, it is little wonder that the *Lancet* paper received a great deal of publicity, even prompting such sensational headlines as "Vitamins Only Take You Closer to Death."

That may be overkill. Although there is no question that the *Lancet* study was well done and has statistical weight, it does leave unanswered questions. For example, what percentage of the subjects had some preexisting condition for which they were taking the supplements? Is it possible that protection against cancer requires that antioxidants be taken for periods longer than twelve years? Maybe these supplements don't protect against cancer, but what about heart disease or other conditions?

So, what do we do with this study? First of all, we do not shoot the messenger. Supplement promoters have already started to accuse *The Lancet* of catering to the pharmaceutical industry by publishing a paper that undermines "natural health products." The argument is that, if people stay healthy by taking vitamins, they will have less use for prescription drugs, so it is therefore in the interest of big pharma to support studies that show supplements to be useless or even dangerous. Bunk. Many pharmaceutical companies make a nice profit from manufacturing supplements.

Unfortunately, there is a growing body of evidence that these supplements are not the saviors they were portrayed to be. A recent meta-analysis of vitamin E studies in the *Journal of General Internal Medicine* concluded that the vitamin is not useful in the prevention or treatment of heart disease. (For more on vitamin E, see "Vitamin E Doesn't Deliver Either," below.) Folic acid and other B vitamins lower blood levels of homocysteine, which has been implicated in heart disease. But a recent double-blind trial showed that patients whose arteries had been opened up by placement of a coronary stent actually did worse if they were given folic acid supplements. And in formula-fed children, vitamin supplementation in the first three months of life was found to be associated with an increased risk of subsequent food allergies.

What are we to make of all this? Jumping to the conclusion that vitamin supplements are dangerous is not justified. Many people in North America have diets that do not provide the recommended daily intake of vitamins, and for them, a one-a-day multivitamin is a good idea. But accumulating evidence suggests that it is better to get our nutrients from food than from supplements. There seems to be an almost magical blend of antioxidants, minerals, and as-yet-unrecognized beneficial compounds in fruits, vegetables, and whole grains, a blend that is not replicated in supplements. Maybe we would be better off by taking the money spent on supplements and putting it toward fresh produce. At least until the next study comes out "proving" that some dietary supplement is the only antidote to death. Like I said at the beginning, science can drive you to despair. But then again, there's this study about how vitamin B6 can increase serotonin levels in the brain and calm you down. . . .

Vitamin E Doesn't Deliver Either

Hey, vitamin E was supposed to be good for the heart. That's what "they" said. And you can count me in that "they." Several years ago, I wrote a pretty optimistic piece on vitamin E, based on some interesting observational studies and some promising laboratory findings. A number of studies had shown that people taking vitamin E had fewer heart attacks and strokes, and investigators had demonstrated that the vitamin had powerful antioxidant effects—at least in the test tube. The only downside seemed to be an anticoagulant effect at high doses. I did have a lingering concern about that, but upon looking into the issue, I concluded that at a dose of 200 IU this was not a concern, and I was quite comfortable suggesting a daily intake of that amount. Couldn't do any harm, and might do some good. I figured that intervention studies, which were just getting started at the time, would nail down the details about the most appropriate dose. They would also eliminate the possibility that vitamin E takers were seeing benefits not because of the supplement, but because they were eating healthier diets and exercising more.

Well, those intervention studies served up a surprise. Study after study showed that people taking vitamin E were no better off, in terms of cardiovascular disease, than those taking a placebo. Even the studies about preventing macular degeneration, a serious eye condition, were disappointing. And then came the analysis of some nineteen vitamin E studies by Dr. Edgar Miller of Johns Hopkins University. In a paper published in the *Annals of Internal Medicine,* Miller reports that, when the results of these studies involving over 136,000 patients are combined, no protection against cardiovascular disease is evident. Then comes the shocker. Not only did the vitamin not protect against disease, but it also appeared to increase mortality from all causes!

This finding startled the millions of people taking vitamin E supplements, and shook supplement producers to the core. The supplement industry is monumental, and profits are huge. Understandably, then, the industry was quick to mobilize and point fingers at what it claimed were shortcomings of Miller's "meta-analysis." Most of the studies involved people who already had cancer, Alzheimer's, or heart disease, and therefore the results would not be expected to apply to a healthy population. While it is certainly correct that the subjects of these studies already had some sort of disease, the inference that the results therefore do not apply to healthy people is not necessarily valid. Indeed, one normally expects an intervention to be most effective in people already burdened with disease. Aspirin, for example, is of great use in preventing heart attacks in people with existing heart disease, but the jury is still out on the effectiveness of healthy people taking aspirin. So if vitamin E offers no help to those who suffer from an ailment, it is unlikely to benefit the healthy.

Another criticism aimed at the study suggests that most of the participants took synthetic, rather than natural, vitamin E. That's true, but it is the synthetic version that is most commonly available, and that's the one people tend to take. There are eight different naturally occurring forms of vitamin E: four different "tocopherol" forms, and four corresponding "tocotrienols." It certainly is possible that supplements composed of a blend of all of these would have a different effect than just alpha-tocopherol, which is the base of most supplements, whether synthetic or natural.

Just as the smoke was clearing from Miller's meta-analysis, the scientific community and the public were stunned by the results of HOPE-TOO, published in March 2005 in the *Journal of the American Medical Association*. The Heart Outcomes Prevention Evaluation Ongoing Outcomes trial (who says scientists don't have imagination when it comes to names?) began in 1993, when more than 9,000 people over the age of fifty-five, with a history of either diabetes or heart disease, were randomly assigned to take either 400 IU of natural source vitamin E, or a placebo. The hope of HOPE was that vitamin E would provide some benefit against heart disease, and perhaps cancer. Alas, such was not to be the case. To the surprise of the researchers who monitored the study, not only did the vitamin prove ineffective, but it actually increased the risk of heart failure by up to 19 percent! Again, we have to remember that the results may have been different in a healthy population. The study used natural source alpha-tocopherol, and, as mentioned above, some critics have suggested that the results would have been different if the supplement reflected the ratio of the four different forms of tocopherol as they occur in the diet. Possibly, but that's conjecture. In any case, alpha-tocopherol is what is commonly found in the supplements people take to ward off disease.

So is vitamin E hopeless? No study can provide a final answer, but the Miller analysis showed no benefit with vitamin E use in 136,000 people, as well as a dose-response relationship in terms of mortality. In general, when an effect, either positive or negative, increases with dose, it usually means that it is real, rather than a statistical artifact. In this case, the researchers noted that risk of premature death began to rise at around a daily dose of 150 IU of vitamin E, and at a dose of 400 IU per day, the risk of dying from any cause becomes about 10 percent higher than for people not taking the vitamin. HOPE-TOO added more fuel to the burning controversy with its finding of a cardiac risk instead of a benefit from vitamin E supplements.

Of course, it is always possible that vitamin takers don't take as much care with their diet and exercise habits because they feel they are protected, but this is not a likely explanation, given the large number of subjects involved in the studies. The fact is that, as more and more high-quality studies about supplements come to light, we begin to see an emerging pattern. While antioxidants undoubtedly play an important role in health, their relative amounts are of essence. More is not necessarily better. Food seems to contain the best balance of these nutrients, and when we flood the body with antioxidants from an outside source, the antioxidant balance is upset to the extent that adverse reactions may occur.

More evidence about the ineffectiveness of vitamin E in preventing cardiovascular disease or cancer comes from the Women's Health Study conducted between 1992 and 2004. Almost 40,000 women over the age of forty-five were given either a placebo or 600 IU of natural source alpha-tocopherol on alternate days. The researchers found no overall benefit for major cardiovascular events or cancer and concluded that the data do not support recommending vitamin E supplementation for the prevention of these diseases in healthy women.

Given the fact that we have no clear indication of benefit from high doses of vitamin E, and that there are suggestions of possible harm, the prudent advice is to avoid high doses. Moderate amounts, though, may provide some benefit. In seniors, 200 IU a day seems to offer some protection against respiratory tract infections, and vitamin E may play a role in preventing Alzheimer's disease. A study of 5,000 elderly people in Utah found that vitamin E supplements protected against Alzheimer's, but only when taken together with vitamin C. This actually has a theoretical rationale: vitamin C can recharge the activity of vitamin E after it has performed its role as an antioxidant. A combination of 400 IU of vitamin E with 500 milligrams of vitamin C was found to be useful. Such amounts are not dangerous. So while vitamin E may not do much for your heart, it may help your memory. You just have to remember to take vitamin C along with it! But once memory problems have appeared, vitamin E is useless. In a study reported in the *New England Journal of Medicine* in 2005, researchers followed 769 older people (ranging in age from fifty-five to ninety years) who had experienced mild cognitive impairment. Some were treated with high doses of vitamin E (2,000 IU per day), and some with a placebo. Vitamin E was ineffective in warding off Alzheimer's disease. The same study showed that donepezil (Aricept) slowed the progression of the disease during the first year, but this benefit was not maintained thereafter.

The jury is still out on the possible benefits of vitamin E in slowing the progress of macular degeneration, an eye disease characterized by the breakdown of cells in an area of the retina known as the macula. In 2001 the Age-Related Eye Disease Study (AREDS) created a great deal of excitement when it revealed that a supplement consisting of 500 milligrams of vitamin C; 400 IU of vitamin E; 15 milligrams of beta-carotene; 80 milligrams of zinc as zinc oxide; and two milligrams of copper as cupric oxide significantly slowed the progress of the disease. The role of vitamin E here is unclear, but in any case, there is no evidence that such supplements may prevent people from getting macular degeneration in the first place.

I must admit that I had greater hopes for vitamin E, based on its antioxidant potential and a few promising epidemiological studies. But placebo-controlled intervention trials have unfortunately not lived up to expectations. Such is the nature of science! It is a self-correcting discipline, and in the long run, science eventually manages to focus in on the truth. The truth may not always be what we wish, but the mark of a scientist is to go according to the results of research, not according to his or her wishes.

The Cold Facts about Vitamin C

Vitamin C and the common cold. You would think that the exact link would be clear by now. After all, it's been more than three decades since Linus Pauling suggested that vitamin C was the answer to this viral misery, and his theory spurred a host of investigations. Had this claim been made by anyone else, the scientific community would have yawned and ignored it. But this was Linus Pauling, a Nobel Prize winner, and perhaps the greatest chemist of the twentieth century. He had contributed so much to our understanding of the chemical bond, the structure of proteins, and the mystery of sickle-cell anemia that many scientists thought his "gut feeling" about vitamin C merited further examination.

Humans are one of just a few species incapable of synthesizing vitamin C, a distinction we share with other primates, guinea pigs, fruit-eating bats, and the red-vented bulbul, a curious bird. Pauling believed that we lost the ability to manufacture the vitamin from components in food somewhere along the evolutionary line, and that we were now paying the penalty. Pauling maintained that the amounts ingested in the diet may be enough to protect us from scurvy, the classic disease brought on by vitamin C deficiency, but that this was not enough for optimal health.

Linus Pauling was one of my chemical heroes when I was growing up, and I remember the excitement I felt the first time I heard him speak at a conference. It was in the early 1970s, and I can still vividly recall the great man strutting around the stage, brandishing a vial holding the amount of vitamin C a goat produces in a day, telling the enraptured audience, "I would trust the biochemistry of a goat over the advice of a doctor." Hmmmm, I thought, that didn't sound very scientific. But this was Linus Pauling. He had to know what he was talking about!

Well, it seems that far more distinguished members of the scientific community than me weren't quite so sure, and decided to put the vitamin C hypothesis to a test. Let's mount some clinical trials, they proposed, and check this out. And how appropriate that was! After all, the first real clinical trial in history involved vitamin C. That goes back all the way to 1747, and the pioneering work of a Scottish ship's surgeon, Dr. James Lind, who is usually credited with having discovered that scurvy, a disease that rotted gums, caused joints to become swollen, robbed the body of energy, and often killed its victims, could be cured with citrus fruit juice. Actually, others had made similar observations before Dr. Lind. When Jacques Cartier's ships became icebound in Quebec in 1536, only three of his 100 men escaped the ravages of the disease. It was then that the native Stadacona people came to his rescue and advised the men to make a tea by boiling the leaves of a tree, probably the white cedar. Rapid recovery was observed after only a couple of doses, but due to poor information transmission in those days, the remedy seems to have been lost. There were other instances of effective scurvy treatment. In the seventeenth century, some East India Company ships carried supplies of lemon juice to ward off the disease. Still, these were isolated cases, and thousands of sailors elsewhere perished from scurvy.

That's when Lind entered the picture. He had undoubtedly heard accounts of treating scurvy with various foods or beverages, and he decided to get to the bottom of the matter. Aboard the HMS *Salisbury*, he selected six pairs of men, and gave them either a daily dose of cider, dilute sulfuric acid, vinegar, seawater, a mish-mash of garlic, mustard seed, and radish root, or two oranges and a lemon. There was also a control group of men with scurvy who got the regular ship's rations. Within days, the two men who were lucky enough to have been put on the citrus diet began to recover. So although Lind was not the first to discover a treatment for scurvy, he certainly was the first to document a "clinical trial" showing the effectiveness of the citrus remedy, which he did in his "Treatise on Scurvy" in 1753.

Still, it wasn't until 1795 that the British navy began to provide a daily supply of lime or lemon juice to all its men.

Dr. Lind would have approved of the clinical trials that various researchers organized to check out Pauling's vitamin C–common cold hypothesis. Over sixty controlled studies have examined the effects of vitamin C supplements on the common cold, in some cases using up to several grams a day. You can pick and choose among these studies to "prove" whatever point you want to make. If you want to show no effect whatsoever, an Australian study of 400 volunteers taking various doses of vitamin C over eighteen months is the example of choice. If you want the opposite result, check out an American study of 463 students over two years. And if you want the true bottom line, here it is. The evidence that vitamin C supplements can prevent the common cold is very sketchy, although there may be an effect in people who have an extremely low dietary intake of the vitamin.

This is corroborated by the findings of two researchers, one Australian, the other Finnish, who examined the published studies to date and subjected them to critical analysis. They looked at some fifty-five placebo-controlled trials that used at least 200 milligrams of vitamin C supplements a day. The results are not spectacular. In people who took vitamin C regularly to prevent colds, there were no fewer colds, but there was a slight reduction in the number of days they experienced symptoms: 14 percent for children; 8 percent for adults. Interestingly, there was a 50 percent reduction in the incidence of colds in marathon runners, skiers, and soldiers exposed to significant cold and or physical stress. There was also some evidence that taking large doses of vitamin C, as much as 8 grams, on the day that cold symptoms appear can shorten the duration of a cold. Indeed, a recent study showed that the synthesis of cytokines, chemicals the body generates to fight viruses, is increased within hours of taking a gram of vitamin C.

Taking vitamin C prior to extreme physical exertion or exposure to cold stress therefore makes sense, as does taking large doses the first day that cold symptoms appear. In any case, the impact of vitamin C supplements on the common cold is not a major one. If it were, we would have seen it conclusively in the studies, and we would not be debating the issue.

A Cancer Treatment on Trial

It was a trial that captured the imagination of the public. Gaston Naessens stood accused of criminal negligence in the death of a cancer patient who had been treated with his elixir 714X instead of undergoing conventional treatment. Not only Naessens, but all of "alternative medicine" seemed to be on trial in that courtroom in Sherbrooke, Quebec, back in 1989. The French-born biologist claimed that his studies had led him to investigate the true cause of cancer, which he found to be the uncontrolled growth cycle of "somatids," living organisms distinct from bacteria and viruses. Mainstream medicine had missed their existence, he said, because they were too small to be seen with an ordinary microscope. But Naessens had devised an instrument, "the somatoscope," that allowed these organisms to be visualized, and the presence and progress of cancer to be monitored. He then theorized that somatids could be returned to a normal state by injection with a drug he had developed. Naessens took great pride in his creation, naming it 714X. The numbers 7 and 14 served to represent his initials (the seventh and fourteenth letters of the alphabet), and X, the twenty-fourth letter, stood for the year of his birth, 1924.

Desperate patients will do desperate things, so it came as no surprise that cancer victims beat a path to Naessens' door. After all, he offered them hope of victory over a disease that too often proved more than a match for radiation and chemotherapy. The fact that the medical establishment maintained there was no scientific basis for the treatment, or indeed for the existence of somatids, did not seem to be a deterrent. Naessens presented himself as a latter-day Galileo who would be vindicated soon, as his successful cures multiplied. Patients were charmed by the man who appeared to be toiling not for money, but for the benefit of mankind, and were seduced by the idea of a maverick researcher with no formal training who found a solution to a problem that had stymied hordes of MDs and PhDs. Had Gaston Naessens really found the answer to the mystery of cancer? Not as far as Angele Langlais was concerned. She had visited Naessens after being diagnosed with breast cancer, and, after hearing his claims, she decided to forgo conventional therapy and rely solely on 714X. When Angele developed kidney pain, Naessens urged her to get an x-ray, and upon viewing it suggested that she was merely suffering from a pinched sciatic nerve, and needed a chiropractor. The painful injections of 714X into lymph nodes in the groin continued, but within fourteen months, Angele Langlais was dead. Throughout this period, Naessens kept comforting her by repeating that she was getting better and would be in her garden by the spring. After her death, the Crown decided to prosecute Naessens for criminal negligence, and tacked on a couple of counts of assault and fraud for injecting patients with a drug that had no established scientific merit.

Prosecutors maintained that 714X, a concoction of camphor, ammonium chloride, ammonium nitrate, sodium chloride, ethanol, and water, had not been properly evaluated as a cancer treatment and, in fact, had no theoretical foundation. Naessens, according to the Crown, had no basis for suggesting that 714X, injected into lymph nodes in the groin daily for twenty-one days, would improve immune function. Furthermore, researchers who had investigated 714X had found no effect on tumors in animals. The defense countered by lining up a number of patients who claimed that they had been helped or cured by Naessens' regimen. The witness with the highest profile was Quebec politician Gérald Godin, who had been battling brain cancer. He told the court that the size of his tumor had been reduced by 60 percent after treatment with 714X, and that "Naessens' potion is curing my cancer." Others also gave emotionally charged accounts of being helped by Naessens after physicians had given up hope.

The jury deliberated for no more than four hours before returning a verdict: "not guilty." Naessens and his followers hailed the victory as a triumph for alternative medicine, and proclaimed that the stage was now set for the truth about 714X to emerge. Why had the jury found Naessens innocent? It couldn't have been because they had been swayed by the scientific evidence on behalf of 714X. There wasn't any. Rather, in the opinion of the six women and five men, patients diagnosed with a dire disease had a right to freely choose their course of treatment, whether this had been shown to be effective or not.

Following the trial, three highly regarded cancer specialists offered to meet with Naessens to examine the evidence for cures he claimed to have documented. Contrary to suggestions by "alternative" practitioners, mainstream oncologists would be delighted to offer any effective treatment to their patients. After all, they, better than anyone else, know the limitations of current therapies. Unfortunately, Drs. Gerald Batist, Jean Latreille, and Jacques Jolivet found no evidence of miraculous cures after examining Naessens' files. In many cases of "success," patients had also undergone conventional treatment, and in others, there was no adequate follow-up. Gérald Godin, it turned out, was actually a patient of Dr. Jolivet, and had undergone surgery, radiation, and chemotherapy. Although he claimed during the trial that he had been cured of cancer by 714X, Godin died of the disease in 1994.

Naessens' 1989 acquittal was interpreted by some as a validation of 714X, and patients besieged the government to make the drug available. Health Canada acquiesced and provided the concoction to physicians under the emergency drug-release program. Over 4,000 patients have apparently been treated. And here we are, twenty-five years after American writer Christopher Bird wrote an impassioned book, *The Life and Trials of Gaston Naessens,* suggesting that the truth about the cure was now primed to come out. He was right. It has. Not a single clinical study of efficacy has been published. Naessens may have meant well, but if 714X were an effective cancer treatment, we would have known it by now.

"THE CURE FOR ALL CANCERS"

Now there's a headline that gets attention! It sure got mine when I saw it on the cover of a book prominently featured in a health food store. This was not a treatment, the book suggested, but a cure! A gift to humanity, the author self-indulgently stated. What had thousands of dedicated researchers missed over the years, I wondered?

Claims of cancer cures are nothing new. In medieval Europe, a live crab would be placed on the body at a site close to a tumor, left there for a while, and then the animal would be removed and killed. Why? Because tumors often bore a physical resemblance to the crab. In fact, our word *cancer* derives from the Latin word for the creature. The idea then was that the tumor would develop some kind of association with the crab, and would somehow be sympathetically destroyed along with the poor crustacean. Judging by the fact that this procedure persisted for a couple of centuries, it must have produced at least some successes. This is not surprising. Sometimes spontaneous remissions do occur, and of course sometimes people feel better, at least temporarily, due to the placebo effect. But even with the popularity of "alternative medicine" today, it is safe to say that anyone who might suggest that crabs can physically withdraw cancer from the body would be regarded as less than sane.

Now, fast-forward to modern times. Imagine that you were suffering from cancer. Imagine that you were told that you could be cured of the disease in just five days by identifying, and then removing, the cause of your cancer. Imagine that all you had to do was buy about thirty-five dollars' worth of parts and build a simple electronic device that would tell you exactly what to do. Imagine that you were instructed to eat a certain food, then squeeze a pimple on your body and place the emerging fluid on the device next to a sealed plastic bag of the same food. Imagine that you were then to connect the contraption to your knuckles by means of two leads and listen to the sound emanating from a little speaker in the apparatus. Now imagine that, by the type of sound emitted, you could determine whether this particular food was a cause of your cancer, and whether it should therefore be eliminated from the diet to ensure a cure. Finally, imagine that you don't have to imagine all this. For indeed the foregoing is the actual scenario being plied to the public in an epic work with the grandiose title "The Cure for All Cancers!"

Hulda Regehr Clark, who surprisingly possesses a PhD in physiology from the University of Minnesota, unabashedly claims to have discovered the secret that has stymied all other scientists. The cause of cancer, she claims, is an intestinal parasite that can escape from the gut and take up residence in a variety of organs. These organs have been weakened by previous exposure to a variety of substances, ranging from mercury in dental fillings and thallium in wheelchairs to wallpaper glue and asbestos in clothes dryers. But the cancer process can only begin if certain other chemicals are concurrently present in the body. Apparently, the greatest culprit is isopropanol, otherwise known as rubbing alcohol. But other solvents, such as methanol or xylene, can also initiate cancer when present together with the parasite. These solvents, according to Clark, are found as contaminants in our foods, drinks, and cosmetics.

The cure for cancer, then, is obvious to the writer. Kill the parasites and avoid all products contaminated with solvents, as well as all chemicals that weaken our organs. These products include shampoos, cold cereals, carpets, stainless steel, porcelain, and toast. Toast, you ask? Of course. Didn't you know that it is contaminated with tungsten from the element in the toaster? How does one go about killing the parasites? A mixture of cloves, black walnut, and wormwood destroys the intestinal flukes, as they are called, and therefore in Clark's words "can cure all cancers." And, naturally, the instrument just described (which Clark calls a "Syncrometer") will determine exactly which foods and other substances must be avoided to result in a cure.

If you want to know whether there is any aluminum in your brain, weakening it and therefore making it more susceptible to disease, the Syncrometer can tell you. According to the detailed instructions, just buy a piece of pork brain, place it on the device next to a piece of aluminum, attach the leads, and listen for "resonance." The pork brain, you see, guides the instrument where to look, and the piece of aluminum tells it what to look for. Similarly, you can use a piece of fish intestine to test for parasites in your colon. How anyone can come up with such a bizarre concept boggles the rational mind. The story would be funny, if the possible consequences were not so sad. Hulda Clark actually uses her Syncrometer to diagnose cancer! She then goes on to cure people of a disease they may never have had.

Clark, in one of many "case histories," describes how a patient had undergone colonoscopy for severe diarrhea and had been pronounced cancer-free by her physician. Yet one of Clark's bizarre tests showed a positive reaction for cancer. "It came as a shock to her that she actually had colon cancer," Clark says. I bet it did. Of course, a week after starting on the antiparasite program, she was pronounced cancer-free. Strangely, the diarrhea was still present. One also wonders about how many people who really may have serious disease resort to this "therapy" at the expense of proven remedies?

But Clark is not completely anti-establishment. She does admit that oncologists are kind, sensitive, compassionate people. But "they have no way of knowing about the true cause of cancer since it has not been published for them. I chose to publish it for you first so that it would come to your attention faster." And publish it she did. Our doctrine of freedom of speech guarantees her right to do so. It guarantees her right to keep publishing. And she does. Now she can cure all diseases, not only cancer. At least that is what she says in *The Cure for All Diseases*. Strangely, though, AIDS doesn't fall under this umbrella, because she has also published *The Cure for HIV and AIDS*. Of course, the doctrine of freedom of speech does not require that that which is stated must be scientifically valid. Free speech emerging from the wrong mouth can be very dangerous!

Praying for Health

Now get a load of this. In July of 2000, Professor Leonard Leibovici of the Rabin Medical Center in Israel gathered the names of patients who had been hospitalized for a bloodstream infection between 1990 and 1996. He randomly divided them into two groups, and gave the names of those in "the experimental group" to a person who said a short prayer for their well-being and recovery. Leibovici then examined the hospital records of the 3,393 patients involved, and published his findings in the *British Medical Journal*. The results were remarkable. Patients who had been prayed for, mind you, four to ten years after their hospitalization, had significantly shorter duration of fever and hospital stays! Leibovici was not put off by this apparently staggering revelation, explaining that we cannot assume time is linear, or that God is limited by time. Intercessory prayer works, it seems, even if it is done years after the fact.

This was not the first study to examine the possibility that patients might benefit from the healing power of prayer even though they did not know that they were being prayed for. The ball got rolling in 1988 when Randolph Byrd, a cardiologist at San Francisco General Hospital, randomly divided 393 heart patients into two groups, one of which then received prayers from participating Christians outside the hospital. The results, published in the *Southern Medical Journal,* showed that the patients who were not being prayed for needed higher doses of medications, and were more likely to suffer complications. Obviously, this study stirred both the scientific and religious communities, and stimulated attempts to reproduce the findings. Some thirteen of twenty-three randomized placebo-controlled trials of "distant healing" carried out since then have apparently shown positive results, albeit amid much controversy. Numerous questions do arise. Does God punish people who are

unlucky enough not to have their names drawn out of a hat? Why does he play pharmacist, tinkering with doses of drugs instead of just healing patients outright? Does the kind of prayer or its intensity matter? What if prayer and prayee are not of the same faith? Does the researcher's belief system affect the objectivity of the studies? These are deep waters, so it certainly is not surprising that most scientists remain unconvinced about the effectiveness of intercessory prayer.

That is certainly not to say that scientists question the value of prayer. Indeed, surveys show that 40 percent of scientists believe in a God who answers prayers. Maybe that's because there is a vast amount of evidence showing that people can benefit from praying. For example, one of the best indicators of survival after heart surgery is the degree of faith patients have in a higher being (not the surgeon). It is also well known that people who attend religious services regularly have lower blood pressure and a lower risk of dying from coronary disease.

Furthermore, the elderly who pray regularly are less likely to be depressed. Maybe divine intervention plays a role, or perhaps, like meditation, praying lowers blood levels of adrenaline and cortisol, naturally occurring compounds that characterize stress.

All of this sounds pretty reasonable. But could prayer affect the outcome of in-vitro fertilization? Sounds far-fetched. But that is exactly what a paper published in the *Journal of Reproductive Medicine* in 2001 implied. The study, with lead author Dr. Rogerio Lobo, chairman of obstetrics and gynecology at Columbia University, showed that infertile women who had been anonymously prayed for became pregnant twice as often as those who had not been prayed for. A truly astonishing finding!

Two hundred women who received in-vitro fertilization in a South Korean hospital had been divided into two groups, with half being prayed for by Christian groups in the US, Australia, and Canada. As is the usual custom, the editor of the journal had sent the paper out to be reviewed by experts in the field of reproduction. While they expressed amazement at the results, they could not find any methodological problems with the study, and recommended it be published. This was exactly the kind of study the press loves, and soon headlines trumpeted the results around the world: "Scientists Show the Value of Prayer!"

Then the skeptics stepped in. They wanted more details. They wanted to know whether the study had the approval of an ethics committee, they wanted to know who had funded it, and they wanted to know why the study design was so convoluted. Some prayer groups, for example, were themselves also being prayed for by other groups to increase the success of their prayers. Bizarre. In any case, answers to these questions were not forthcoming. One of the authors of the study had left Columbia and would not respond to inquiries, and it turned out that the "lead" author, Dr. Lobo, only learned of the study six to twelve months after it was completed, and only provided "editorial review." Basically, he took the word of the other authors about what the study had shown.

Now we come to another problem. Namely, that the third author of this remarkable study, Daniel Wirth, had no connection to Columbia, but had previously published various articles about supernatural healing. He holds a degree in parapsychology. And in 2002, after the prayer study, he was indicted on thirteen counts of mail fraud and twelve counts of interstate transportation of stolen money. All of this casts great suspicion on the prayer study, and one even wonders if the study was ever actually done. Professor Leibovici may have some thoughts on this matter. It seems that a couple of years before his "research" about intercessory prayer, he had written another article in the same journal about the pitfalls of "alternative medicine." Pray tell, could the study about retroactive prayer have been a tongue-in-cheek admonition to researchers who would dare to venture into areas not amenable to scientific research?

GERMS, GERMS EVERYWHERE

Don't look now, but you are full of germs. Billions and billions and billions of them. Bacteria, viruses, and fungi of all kinds are at this moment cavorting in your saliva, on your skin, and in your bowels. Just your intestines contain about a thousand billion germs per gram of matter; a number greater than the total number of cells in your body. And if you really want to be impressed, just consider that the combined weight of germs in the world is greater than the combined weight of all plants and animals. It's hard to imagine such an astounding presence, because you can look for germs and not see a single one. Not without a microscope, anyway. But you can imagine the shock and awe when, in the seventeenth century, Antonie van

Leeuwenhoek managed to build a microscope that magnified some 200 times! In his own saliva, he saw "many very little living animalcules, very prettily moving. The biggest sort had a very strong and swift motion, and shot through the spittle like a pike does through the water." This was the first description of the antics of bacteria! Van Leeuwenhoek had no idea of the role his "animalcules" played in life, and would undoubtedly have been shocked to learn that diseases attributed to microbes are the number-one killer of people in the world.

Possibly, Girolamo Fracastoro, an Italian physician who lived 100 years before van Leeuwenhoek, would have interpreted the observation correctly. He had theorized that diseases were transmitted by "seminaria," tiny particles that were way too small to be seen. But Fracastoro's theory and van Leeuwenhoek's observations would not coalesce until the brilliant French chemist Louis Pasteur became interested in why French wines sometimes soured. He investigated numerous wines with his microscope, and found that when the wine was up to snuff, the only microbes present were yeasts from the grape, but when it became cloudy and foul smelling, there were various bacteria present. Where did they come from? Pasteur quickly proved that the source of bacteria was the air. He devised fermentation vessels that prevented contact between the wine and the air and showed that the wine did not sour. Now he began to wonder whether, if microbes could make wine sick, they could do the same to people.

The fact that certain diseases were contagious had long been evident, but nobody knew exactly how they were transmitted. Some clues had, however, been uncovered. Anthrax, an animal disease, could be transmitted if blood from a sick animal was injected into a healthy one, and it was assumed that some chemical in the blood was responsible. Pasteur did not believe this was so. He had observed tiny rod-shaped bacteria in the blood and thought these were the culprits. To prove his point, he devised a clever experiment. Pasteur took a drop of fresh blood from a sheep that had just died from anthrax, and placed it in a flask with a culture medium in which bacteria could multiply. After allowing the microbes to proliferate, he transferred a drop to a second flask, diluted it with water, and again allowed the microbes to multiply. He did this a dozen times, until he was sure that none of the original blood present remained. Still, an animal injected with a sample from the twelfth flask died of anthrax! The disease was not caused by blood, but rather by the bacteria it harbored.

Pasteur's idea that specific germs caused specific diseases was not immediately accepted. One disbelieving surgeon even challenged him to a duel! Today we know that Pasteur was correct, and that germs are responsible for a great deal of misery. *Clostridium difficile,* so called because at first it was very difficult to culture in the laboratory, is a particularly nasty microbe. It can cause terrible diarrhea, and is usually acquired in a hospital after a patient has had antibiotic therapy. The hundreds of varieties of bacteria that live in our digestive tract usually do so in harmony; they compete for food and keep each other in check. But antibiotic therapy can wipe out some of the "good" bacteria and allow *C. diff* to multiply and produce toxins that cause diarrhea or, in severe cases, even permanently damage the large bowel. Infection with this bacterium rarely causes any significant problem for people who are in good health and who have not been on antibiotics.

Usually just stopping antibiotic therapy, if possible, resolves the problem. If not, oral metronidazole or vancomycin is tried. A promising treatment makes use of oral doses of lactobacillus GG ("Culturelle"), a "good" bacterium that can survive the journey through the stomach and suppresses the growth of disease-causing bugs like *C. diff*. Another fascinating possibility is the insertion of fecal bacteria from a healthy donor into the bowel of a patient with hopes of reestablishing a balance between the good and bad bacteria.

Without a doubt, though, the best way to deal with *C. diff* is through proper hygiene. The bacterium forms "spores" that can live outside the body for months. Alcohol gels are ineffective against spores, but washing the hands with soap for thirty seconds gets rid of them. Cleaning floors, sinks, and toilets with bleach also destroys the spores. Remember that the ten worst sources of contagion are our fingers. If only hands were washed properly, and repeatedly, the *Clostridium difficile* problem would not be so difficult to deal with.

Music and the Savage Breast

I'm not sure what a "savage breast" is, but I'd be willing to bet that music can soothe it. I doubt that it can do much about the hardness of rocks, but I wouldn't rule out an effect on vegetation. Of course I'm referring to a 300-year-old quote by British playwright William Congreve, which in its entirety reads "Music has charms to soothe a savage breast, / To soften rocks, or bend a knotted oak." Yes, the correct quote is "breast," not "beast," although the latter misquote is probably more common than the original. And savage beasts probably can be soothed by music.

Cows, except for some mad ones, may not exactly be savage beasts, but it seems they are affected by music. A herd in Indiana increased its milk production by more than 5 percent when a Beethoven symphony was piped into the barn. Country music had no effect, but when heavy metal was played, the animals didn't even want to enter their stalls. And when they did, they were not interested in giving milk. Production decreased by 6 percent. I don't know of any oak trees that have been bent by music, but back in the 1970s Dorothy Retallack made it onto the CBS evening news with her research about the musical tastes of plants. Petunias apparently leaned toward a tape player in response to Ravi Shankar's sitar music but attempted to escape when rock and roll was played. An Illinois botanist even found that corn grew taller when exposed to Gershwin's "Rhapsody in Blue." I wonder if it would grow as tall as an elephant's eye if it listened to "Oklahoma."

Now let's get down to the kind of savage beasts—and breasts, I suppose—that interest us the most. Those of the human variety. There is no doubt that we are soothed by music. If you have to undergo a colonoscopy, according to recent research, you'll be less apprehensive and more cooperative if you can listen to your favorite music. Even the doctor may perform better. When surgeons were given the task of counting backward, subtracting the number thirteen each time, they had a lower pulse rate and lower blood pressure and did the arithmetic faster when listening to music. Japanese researchers have shown that levels of cortisol, a stress hormone, decrease in response to music, but testosterone levels show a gender bias. In men, music seems to lower testosterone, but it increases it in women. So, after dinner, gentlemen may want to queue up Ravel's "Bolero" for their partner while they leave the room to do the dishes. They can then come back for dessert and put on a Mozart sonata. Studies have shown that fewer calories are consumed when such soothing arrangements are played.

Ah, those Mozart sonatas! They can do more than cut down on calories. That's what Gordon Shaw, a physicist, and Frances Rauscher, an expert in cognitive development, at the University of California found back in 1993, when they designed an experiment to study the effect of Mozart's music on the brain. Rauscher was a former concert cellist who realized that music could change people's moods and wondered if it could alter their thought processes as well. Three dozen students were asked to listen to the first ten minutes of Mozart's "Sonata for Two Pianos in D Major" before being given a paper folding and cutting test. The students who listened to Mozart performed better than the ones who took the test after listening to a relaxation tape or after a period of silence. These results captured the imagination of the lay press and spawned an industry. Articles appeared about IQs being increased after listening to Mozart, diseases being cured, and children's ability to reason being increased. None of these was supported by solid research. In fact, nobody has been able to duplicate the original work of Shaw and Rauscher. Critics suggested that music improves mood and that nobody should be surprised that people in a good mood perform better on certain tests. In response to a study that seemed to show that kids who took music lessons had higher IQs, the critics maintained that it was probably due to the fact that well-educated parents are more likely to send children for lessons, and children from such parents are more likely to have higher IQs, partly due to heredity.

But now it seems the critics may have been a bit too critical. Rauscher and Hing Hua Li, a geneticist at Stanford University, have come up with some new findings. They had rats listen either to Mozart, or to "white noise." The Mozarted rats performed better in solving a maze, and more interestingly, showed anatomical changes in their brains. They produced more compounds that are known to forge connections between nerve cells, connections critical to enhanced brain activity. It seems that musical stimulation really may have measurable neurochemical effects. Some researchers even suggest that Mozart's compositions may have a special quality that mimics the rhythmic cycles in the human brain. Preliminary research does indeed show that Alzheimer's patients perform better on some spatial tests after listening to a Mozart sonata, and that premature babies can be soothed by such music. Unborn babies have been seen to "dance" in the womb in response to Mozart's compositions. And now researchers at the University of Hong Kong have shown that children who receive musical training do better on verbal memory tests.

Personally, I'd like to see some research on how Broadway tunes affect us. For me, listening to Andrew Lloyd Webber's "Music of the Night" or "Love Changes Everything" is positively therapeutic. Albert Schweitzer, I think, would have agreed: "There are two means of refuge from the miseries of life: music and cats," the famed doctor maintained. The music, I'll buy; the cats, I'm less sure about. Unless one of them sings "Memory" on a Broadway stage.

A STINK ABOUT ANTIPERSPIRANTS

Mum was the word back in 1888 when it came to tackling armpit fragrance. A zinc oxide–based cream, "Mum," was introduced to control the growth of bacteria that cause underarm odor. Its deodorant descendants today mostly contain alcohol or triclosan as antibacterial agents, with some pleasant masking fragrance thrown in. In 1902, Mum was joined by "Everyday," the first antiperspirant. The active ingredient was aluminum chloride, which reacted with moisture to form gelatinous aluminum hydroxide, which in turn blocked sweat gland ducts. Unfortunately, it also irritated the skin and rotted clothing.

By 1947 it had been replaced by the less acidic aluminum chlorhydrate, which is still the mainstay of antiperspirants. Many current products are combinations of antiperspirants and deodorants. And "mum" is certainly not the word about them. The use of these products is being vociferously linked to breast cancer!

Claims of a connection between underarm cosmetics and breast cancer have long been circulating on the Internet. Experts have pretty much dismissed them for lack of evidence that any of the ingredients is carcinogenic, and because the suggestion that antiperspirants do harm by preventing the body from eliminating toxins is absurd. But a paper by Dr. Philippa Darbre of the University of Reading in England in 2004 proposed a novel mechanism by which underarm cosmetics may trigger cancer. The culprits being fingered this time are parabens, pre-servatives commonly used in a large variety of consumer prod-ucts. Darbre examined twenty tumors excised from breast cancer patients and found parabens in eighteen of these. She claims that these chemicals have estrogenic activity and there-fore could be expected to be involved in breast cancer. At first, it sounds like she has made a pretty good case. But wait a minute!

The mere presence of parabens in tumor tissue does not mean they caused the tumor. It is possible that tumors may retain parabens more efficiently than other tissues, and, more signifi-cantly, it is possible that comparable levels exist in healthy tissue. These questions cannot be answered because the researchers ran no controls; they did not investigate whether or not these preservatives were also present in healthy breast tissue. It's a good bet they are. Parabens are commonly used in a variety of foods and cosmetics, and would be expected to show up every-where. In any case, even if parabens turn out to be cancer culprits, there is no way to know that they found their way into breast cells from deodorants or antiperspirants rather than from a myriad of other products. Indeed, today, hardly any underarm cosmetics contain parabens.

Dr. Darbre did not investigate what sort of chemotherapy her now-famous twenty patients may have undergone. Since some drugs are preserved with parabens, this certainly would be relevant information. Nor did she determine whether the women had used antiperspirants or not. Surely, this would be an obvious question to raise. Indeed, scientists at the Fred Hutchinson Research Center in Seattle did raise it in 2002, and published their results in *The Journal of the National Cancer Institute*. They investigated 813 women who had been diagnosed with breast cancer, and compared their underarm cosmetic use to 793 randomly selected controls who were free of the disease. There was no link found between the use of antiperspirants or deodorants and breast cancer. This was a major epidemiological study, yet it was not referenced at all by Philippa Darbre, a case of scientific negligence.

Dr. Darbre also reports that tumors are more frequently found in the left breast, and suggests that this may be due to more women being right handed. These ladies would apply more antiperspirant or deodorant under their left arm. An interesting theory, but totally unsubstantiated. Nobody has studied whether or not right-handed people apply deodorants under their left arms more vigorously. Darbre also points out that there has been an increase in breast cancer rates since the 1970s and that this increase parallels the use of underarm cosmetics. Of course, such a relationship does not prove cause and effect. Most researchers believe the increase is due to an increase in obesity and to women having children later in life. Basically, there is room here for further investigation, but contrary to the beliefs of the apocalyptic types out there, Darbre did not prove that antiperspirants or deodorants are a cause of breast cancer.

Despite the preliminary nature of her study, Dr. Darbre has energetically voiced her views in the media. She was no doubt pleased to see a new study by Dr. Kris McGrath of Northwestern University, who surveyed some 400 breast cancer patients and found that those who shaved their armpits at least three times a week and applied deodorant at least twice a week were generally about fifteen years younger when they were diagnosed with the disease than women who did neither. Deodorant use or shaving alone was not linked to risk; only the combo was. McGrath suggests that nicks during shaving may pave the way for chemicals to enter the body. This is a rather tenuous argument, given that they would end up in the bloodstream and not necessarily in breast tissue. As with Darbre's experiment, there was no control group. We don't know whether or not healthy women have a different pattern of shaving and deodorant use than breast cancer victims. Without this info, we can just as readily conclude that young women shave more frequently and use more deodorant than older ladies.

So where does it all leave us? It is comforting to know that the British, who use deodorants more than Italians, do not have a higher breast cancer rate. The Japanese, who spray and roll as often as Americans, have one of the lowest rates in the world. But those of you who have decided to go "au naturel" based on these studies can take solace in the fact that some researchers have found that underarm aromas may have an aphrodisiac effect. So go ahead and give l'air d'armpit a try. But don't tell anyone I said this. Mum's the word.

The Dangers of Betel Beauties and Fruit-Eating Bats

In Taiwan, they call them "betel beauties." The attractive young ladies can be found on many a street corner, enticing customers with their charming smiles, miniskirts, and nuts. Betel nuts. Several hundred million people in South Asia chew these regularly, but nowhere has the practice attracted as much scientific or political attention as in Taiwan. The betel nuts, which grow on the *Areca catechu* palm tree, are usually mixed with some lime (calcium hydroxide), wrapped in the leaves of the pepper plant, and placed inside the cheek for sucking and chewing. Why? Because compounds in the nut lead to a warm sensation, a general sense of well-being, heightened alertness, and apparently an increased capacity to do work. Sounds good, right? So you know there must be a "but" coming. And there is. Studies have shown that chewing betel is linked with an increased risk of heart disease, diabetes, and asthma. The greatest concern, though, is the strong association between chewing betel and cancer of the mouth. Indeed, in some South Asian countries, oral cancer is the most common malignancy.

The main psychoactive compound in betel is thought to be the alkaloid, arecoline, which mimics the action of acetylcholine, a neurotransmitter involved in the functioning of the nervous system. It is well known that such acetylcholine-like activity can lead to hallucinations, along with a bunch of side effects. Excessive salivation, for one. That's why betel nut chewers constantly defile sidewalks and roadways with their spittle. It's not a pretty sight. The nut stains the mouth and saliva a deep red color, and high-traffic streets can be readily identified by the plethora of crimson spots on the pavement.

Arecoline, however, does not explain all the physiological effects seen in betel chewers. Plasma levels of norepinephrine and epinephrine (adrenaline), two hormones secreted by the adrenal glands, are also elevated, probably accounting for the increased heart rate and skin temperature. Specific polyphenols found in the pepper leaves used to make the betel wad may be responsible for this effect. Then there is also the effect on the eyes. Apparently, male drivers become so distracted by the scantily clad betel nut girls that they have been known to wrap their vehicles around telephone poles. On top of all of this, there is an environmental issue. Because growing betel palms is very profitable, many farmers have replaced traditional crops, like rice, with betel palms, causing severe erosion of hillsides. It is little surprise then that Asian countries, with Taiwan in the lead, have mounted aggressive anti-betel programs. Taoyuan County, near Taipei, has issued an edict requiring betel girls to fully cover the enticing parts of their anatomy. The ladies, of course, are not happy about this, claiming that dressing sexy increases their earnings.

Inhabitants of the island of Guam face a different kind of nutty problem. But this part of our story starts not in Guam, but in New York, at Yankee Stadium. On July 4, 1939, baseball fans became teary-eyed when they listened to Lou Gehrig, their beloved first baseman, announce his forced retirement from the game. "Fans, for the past two weeks you have been reading about the bad break I got. Yet today, I consider myself the luckiest man on the face of the earth." The bad break that Gehrig referred to was amyotrophic lateral sclerosis (ALS), a terrible progressive motor neuron disease that usually leads to death within a few years of diagnosis.

Chances are that, while Lou Gehrig certainly knew a lot about bats, he didn't dream that they might one day yield a clue about the devastating disease that has come to bear his name. Well, we're not talking about the kind of bats that Gehrig was familiar with. We're talking about the fruit-eating bats found on the island of Guam. Starting in the early 1900s, the Chamorro, a native people on the island of Guam, began to show a frightening incidence of a neurological disease that slowly robbed people of the use of their muscles and eventually killed them. By about 1940, it had become the leading cause of death among adult Chamorros. A sudden and dramatic rise in the incidence of a disease is often caused by some sort of environmental factor, such as a toxin in the food supply. One of the main features of the Chamorro diet is flour made from the nuts of the cycad tree. When researchers isolated a compound called beta-methylamino-L-alanine from cycad nuts and found that it was a neurotoxin, they believed they had found the answer to the mystery. But there was a troublesome question. Why did the disease escalate only around 1904? Cycads had always been a part of the Chamorro diet. A further complication turned up when they found that the toxin was destroyed when the nuts were ground into flour. So if the cycad toxin was indeed the culprit, it was getting into the Chamorro's body by some other route.

Oliver Sacks, the famed neurologist and writer, provided an idea. He had discovered that the Chamorros had a particular penchant for fruit-eating bats, which they would boil in coconut milk and devour completely. Catching the bats, though, was quite a challenge, so they were only served on special occasions. At least until the Americans acquired Guam after the Spanish American War. Then guns became much more readily available, and bats appeared more frequently on dinner tables. And guess what! The favorite food of the fruit-eating bat is the cycad nut.

So the theory is that, as natives started to eat more bats, they were exposed to more of the toxin that the bats had concentrated in their bodies. But by the 1940s, the bats had been hunted almost to extinction, and the disease rate dropped. Certainly an interesting theory, and one that merits further investigation, especially after researchers found that a bat can harbor as much beta-methylamino-L-alanine (BMAA) as contained in a ton of processed cycad flour. It's a safe bet that Lou Gehrig never dined on fruit-eating bats, but the evidence gathered in Guam suggests that there may be other toxins in our environment responsible for Lou Gehrig's disease. Gehrig himself passed away just two years after his famous "luckiest man" speech.

BMAA may play a role in another devastating disease as well. The neurotoxin has been found in the brains of nine Canadians who died from Alzheimer's disease. They had certainly never consumed fruit-eating bats, so how did they acquire the toxin? Cyanobacteria, commonly found in lakes, oceans, and the soil, are known to produce BMAA, which can then build up in the food chain. A great deal of further research is needed to clarify any potential link between environmental toxins and Alzheimer's disease, but an appropriate start would be an analysis of the BMAA content of drinking water. Just what we need: another toxin to worry about.

MURDER BY TOXIN

When I was in London, I just had to go to Waterloo station. I wanted to stand at the bus stop right in front of it. I wanted to see with my own eyes the famous spot where one of the most ingenious chemical crimes in history was perpetrated. For it was right there, in full daylight, on Thursday, September 7, 1978, that a Bulgarian émigré by the name of Georgi Markov was assassinated by the Bulgarian secret police. And what an assassination it was! No guns, no grenades, no knives. Just an umbrella! But it wasn't just any umbrella. This was an especially designed umbrella, with a spring-loaded device built to deliver a pinhead-sized pellet loaded with one of the most potent poisons known to mankind. But let's set the scene first.

Markov had become disillusioned with the communist ideology in his native Bulgaria because of the corruption he saw in the higher officials of the government. So he defected to Italy in 1968, and then moved to London in 1971. In London, he joined the Bulgarian service of the BBC. He began to vigorously attack the Bulgarian government over Radio Free Europe. Obviously, the Bulgarians were not happy with this, and threatened to eliminate him if his attacks did not stop. Markov was undeterred, at least until that fateful September day. While waiting at that bus stop across from Waterloo station, Markov felt a jab in his thigh; he turned around to see a man mutter a quick apology as he jumped into a taxi. As Markov later recalled, the man was clutching an umbrella. The Bulgarian expatriate soon started to feel ill, and had to be admitted to hospital the next day. He had a high fever and severe abdominal pains. Within a couple of days, he was dead.

Since he had told his doctors about the jab in his thigh, Markov's body was carefully examined. A perforated tiny metallic pellet was discovered exactly where he had indicated that he had been stabbed. A residue of a poison known as ricin was found in the pellet! Ricin is a protein found in the seeds of the castor bean plant, the same plant that yields castor oil. Castor oil sure conjures up some distasteful memories for me, and probably does so for many of you as well. It's a classic old remedy for constipation, and I shudder to recall how I was forced to take it when my mother diagnosed that I was, let us say, in need of relief. What a horrible taste it had. But it did get the job done. So how come I'm here to tell the tale, given that ricin is one of the most toxic materials that exists? It's because—fortunately—ricin is not soluble in oil. When the oil is extracted from the castor beans, it is washed with water, which completely removes all traces of the toxin.

While there is no danger in consuming castor oil, eating the whole seeds can be deadly. Such poisonings have occurred. Since the seeds are very pretty, they are sometimes used to make ornaments. In fact, in Mexico, castor beans are used to make jewelry for tourists. There can be serious reactions from eating a bean, or even from crushing one in the hand and then putting that hand in the mouth. The symptoms are bloody diarrhea, vomiting, and then shock. So it is obviously not a good idea to eat your Mexican bean jewelry. At least not unless you've cooked it; heat does destroy the ricin. But if ricin gets into the bloodstream, the prognosis is not good. There is no antidote to ricin poisoning, as Georgi Markov found out.

Now talk about coincidence. As I was imagining the umbrella murder while standing at that bus stop, I was startled by the cry of a newspaper vendor across the street. "Professor Plots Deadly Poisoning!" he screamed. "Read all about it!" And I did. What a story it turned out to be.

Professor Simon Wilson, as the article revealed, had been a "leading mathematician" at Manchester University for over thirty-five years. He had never run afoul of the law, at least not until he attempted to pick up a parcel that he had ordered from a biochemical firm under a fictitious name. When he attempted to claim the package, the police were waiting. They had been alerted by the company because the substance that had been ordered was palytoxin, a poison first identified in a species of Pacific coral in 1981, and one that had displaced ricin as the "most potent toxin known to mankind." The coral is found only in a small tidal pool on the island of Maui in Hawaii, and came to researchers' attention because the locals once used it to poison their spear tips. It is available today for research purposes. The company that sells it became alarmed when Wilson ordered enough of the toxin to kill at least 500 people, and decided to notify the police. As it turned out, this was a wise decision.

Simon Wilson had indeed planned murder. Not five hundred murders, though. Only one: his own. Dr. Wilson had planned to commit suicide. He had a pretty good reason, too. His wife had moved her lover into the couple's home. There is, of course, some history here. The Wilsons' marriage had been going downhill ever since they suffered the tragic loss of a child to leukemia in 1975. Communication between husband and wife had basically ceased, and Mrs. Wilson eventually found solace in the arms of a mutual friend. When the friend came down with pneumonia, Mrs. Wilson asked him to move in so she could take care of him. Wilson suspected an affair, but never confronted his wife. When she finally decided to bare her soul and tell her husband about the affair, Dr. Wilson also made a decision. He decided that he did not want to live anymore.

Wilson thought about killing himself with cyanide or with chloroform, both of which the police found in the house. But then, it seems, he read about the potency of palytoxin and decided it was the right chemical for the job. And it would have been, had he not been caught. Dr. Wilson was charged with attempting to acquire a poison, and subsequently appeared in court. The judge, however, accepted his explanation that he had planned no mass murder, and put him on twelve months' probation.

By the time I got through reading this captivating account, a crowd had gathered at the bus stop. I tried to visualize how that famous crime could have been carried out when I was jostled aside by a man with an umbrella who had just hailed a taxi. He quickly scampered in. Was he trying to make a quick getaway? He sure was. It had started to rain.

THE PRICKLY PROBLEM OF ACUPUNCTURE

The rat didn't squirm much as the needles were carefully inserted. There was no pain. Stopwatch in hand, the researcher then focused a heat gun on the animal's rear and carefully noted the time required for the tail to "flick" by reflex action. Other rats, some needled and some not, were also subjected to the same procedure. The conclusion was that acupuncture significantly delayed the onset of pain!

Proponents of acupuncture often quote this study as proof that the 2,000-year-old Chinese technique of jabbing needles into the body to treat medical conditions has a scientific basis. After all, they claim, the effect cannot be in the rats' minds, as they have none! But a rat is not a miniature person. So what's the story about acupuncture for humans? Let's take a stab at that question. We'll begin with a bit of history.

The Western world was alerted to the practice of acupuncture in 1972, when Richard Nixon visited China. A *New York Times* reporter who accompanied the president had to undergo an emergency appendectomy, and received acupuncture therapy for postoperative pain. He was so impressed by the unusual nature of this treatment that he related his experience in the *Times*. The story then took on a life of its own, and rumors mounted upon rumors.

"Did you hear about the guy who had his appendix removed in China without any anesthetic?" buzzed the incredulous. "Those clever Chinese just stuck a few needles into him, and he felt nothing!" But the truth was that the surgery had been carried out with regular anesthesia, and acupuncture had only been used in an attempt to dull the reporter's pain after the operation. Nevertheless, the appetite of the American public had been whetted by the notion of such an apparently simple solution to the problem of pain. The government assembled an

Acupuncture Study Group, composed of many notable physicians and researchers, to study the Chinese experience with the technique. In 1974, off they went to the mysterious East to learn how to rid American pain sufferers of their reliance on painkilling medications.

What they found in China did not exactly live up to the advance billing. Acupuncture, as it turned out, was not widely used as an anesthetic procedure. Even when it was used, it was almost always in combination with barbiturate-type sedatives and the painkiller Demerol. The glorification of acupuncture, it seems, was based less on science and more on Maoist propaganda designed to fuel the Cultural Revolution and justify a reduced reliance on Western medicine.

The American doctors investigating the Chinese accounts discovered that the claimed benefits for acupuncture were highly suspect. Parkinson's patients who were being rehabilitated with acupuncture showed no objective improvement. Neither did patients being treated for deafness or head injuries. But just the fact that American physicians were seriously examining acupuncture caused a flurry of interest back home, and soon half-baked acupuncturists were cropping up in every corner, to treat every conceivable disease. The zanier ones were "curing" conditions ranging from multiple-chemical sensitivity in humans to listlessness in goldfish and neuroses in birds. The most amusing spin-off was "earth acupuncture," which involves hammering wooden stakes into the ground to make the soil more fertile by "altering the earth's natural flow of energy." I think the only fertile thing here is the imagination of the proponents of "earth acupuncture." And this brings up an interesting question. What role does the imagination play in the use of acupuncture to treat disease?

According to the traditional Chinese view, the mind is not involved. Health is based upon yin and yang, opposite forces in

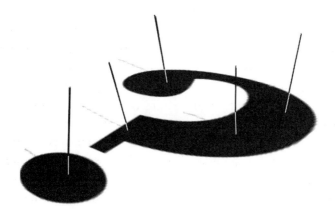

the body that must be in balance to maintain health. When yin-yang disharmony occurs, balance can be restored by stimulating channels in the body, which are called "meridians" and supposedly serve as energy conduits to various organs. Acupuncture points are said to lie along these meridians. This explanation, however, cannot be supported by anatomy. There are no identifiable meridians in the human body. Still, the fact that the explanation may be incorrect does not negate the possibility that acupuncture can deliver the goods. The only way to determine if it really works is through controlled studies.

"Placebo control" is the gold standard for modern clinical trials. We know that people respond to sugar pills, and even to fake surgery, roughly 40 percent of the time, a factor that must be taken into account if a study is to be reliable. Designing a placebo-controlled trial of acupuncture, though, presents some obvious problems. What do you use for a placebo? Researchers at Heidelberg University in Germany came up with a very clever idea. They designed a device that could be used in a "sham acupuncture" procedure. Essentially, the idea is modeled on the classic "collapsing knife" effect used on the stage. The blade is gimmicked to retract into the handle as the victim is stabbed, creating the illusion that the knife has plunged into the flesh to the hilt. Similarly, the sham acupuncture device is equipped with a needle that contacts the skin and then moves up into the handle as the device continues its forward motion. It seems as if the needle has been inserted into the skin. Testing on volunteers revealed they did not suspect any absence of actual penetration.

Over 200 women undergoing breast or gynecologic surgery were enrolled in the study, with half receiving the sham treatment, and half having real needles inserted into an acupuncture point on their forearm. Stimulation of this "Pericardium 6" point is traditionally believed to control nausea. Unfortunately, in this experiment, it didn't. There was no significant difference in postoperative nausea between the two groups, although the patients who had undergone acupuncture did have less of a tendency to vomit. Acupuncturists argue that the results would have been different if several more points had been stimulated. Perhaps.

Since the 1970s, over 500 studies of acupuncture have been published in peer-reviewed scientific journals. Conditions ranging from asthma, drug addiction, and weight reduction to smoking cessation, stroke, and tinnitus (ringing in the ear) have been examined. One would think that so many studies should be able to clarify what acupuncture can and cannot do. Alas, such is not the case. There is much contradictory and inconclusive evidence, as is evident in an excellent recent review published in the *Annals of Internal Medicine,* a highly respected journal. Ted Kaptchuck of Harvard Medical School surveyed the literature for the best controlled trials, and categorized these in terms of the conditions they attempted to treat and the results they found. Kaptchuk holds a degree in Oriental medicine and uses acupuncture himself, so he certainly is not biased against the technique.

"Good" evidence was found for alleviation of dental pain, reduced vomiting after surgery or chemotherapy, and nausea associated with pregnancy. For chronic pain, headaches, back pain, asthma, and smoking cessation, the evidence was contradictory, while for addiction and tinnitus it was negative. A British study has shown that weekly acupuncture, in combination with appropriate medication, can do more to reduce the number of days migraine patients suffer a headache than medication alone. How do we explain the contradictory evidence for the treatment of pain? How can it be that some studies show significant positive results but others come up empty? As with other treatments, the expertise of the therapist is undoubtedly important.

Physicians who are trained in acupuncture are the most likely to select the best candidates for treatment, and of course are also adept at diagnosing conditions that require a more orthodox approach. In other words, you have to know when to reach for the antibiotic or the scalpel, and when to reach for the acupuncture needles. The specific nature of the acupuncture technique is also important. Many of the positive results are seen when electrical stimulation is applied through the needles.

Given that acupuncture can in some cases produce positive results, we are saddled with the question of "how?" One explanation involves the body's production of painkilling substances when stimulated by acupuncture. These "endorphins" are also produced in times of stress, and have been invoked to explain why soldiers often don't feel their wounds until after the heat of the battle. Some studies have shown that naloxone, a drug that blocks the activity of the endorphins, can also negate the benefits of acupuncture. Professor Ronald Melzack of McGill University, regarded as one of the top experts on pain in the world, points out that acupuncture is not some magical process, but rather just one of many methods that can relieve pain through sensory hyperstimulation. Flood the body with sensory input, he suggests, and there will be pain relief. And I believe him. Many years ago, I had the pleasure of listening to one of Professor Melzack's lectures in which he described how rubbing an ice cube on the skin between the thumb and forefinger can alleviate a toothache. Hmmmm, skeptical me thought at the time. Then it happened. I woke up one night with a terrible toothache. I took aspirin, I applied oil of cloves to my gums, but nothing helped. Out of desperation, I reached for an ice cube and (feeling somewhat foolish) began to massage my hand. Within minutes, there was blessed relief! Endorphin release? Opening up of "qi" channels? Placebo effect? Who cares! The pain was gone.

The answer to what happened may eventually come from functional magnetic resonance imaging (fMRI) studies of the brain. Scans of this type have already revealed that when subjects' fingers are immersed in hot water, certain areas of the brain are activated, and that the activity is reduced with acupuncture. But whether the needles are inserted along "meridians," or elsewhere, seems to make no difference.

What are we to conclude from all this? That acupuncture works in the hands of some practitioners, for some conditions, for some patients, some of the time. This may not sound too optimistic, but the truth is that the same statement can also be made about a number of "orthodox" medical treatments. The best bets for acupuncture seem to be in the areas of dental pain, nausea, and migraine. There may be some help with osteoarthritis, although the benefits fade with time. Risks are minor; they are those associated with any needle use. So, as you can see, the facts don't pop the acupuncture balloon, but perhaps they do deflate it somewhat.

REEFER MADNESS

What's the worst movie ever made? Let me put in a plug for the 1936 flick *Reefer Madness*. Disguised as a movie with a plot, this was actually a propaganda film intended to highlight the evils of marijuana smoking. "Women Cry for It, Men Die for It," screamed the promotional posters. Moviegoers would witness "drug-crazed abandon and the soul-destroying effects of killer marijuana." The movie delivers some of the worst acting you'll ever see, along with exaggerated allegations about the effects of smoking marijuana. The thin plot focuses on a pair of upstanding teenagers who fall into the clutches of a dastardly gang bent on converting them into marijuana addicts. It takes only one joint to get the kids hooked, and after that, it's a straight descent into hell. Along the way, there's illicit sex, hallucinations, murder, suicide, and ghastly dialogue. If marijuana makes people talk like that, it is a dangerous substance indeed.

The 1930s featured some of the strongest anti-drug rhetoric in history. Government pamphlets warned teenagers about friendly strangers who might put the killer drug marijuana in their coffee, and described how insanity and death lurked in this "narcotic." You could "grow enough marijuana in a window box to drive the whole population of the United States, stark, staring, raving mad," declared an article in the widely circulated Hearst newspapers. The writer went on to ask the rhetorical question, "heroin, cocaine, morphine, marijuana, opium—what does it matter which it is? One horror is no worse than another." It didn't seem to matter that there was absolutely no evidence to back up these preposterous claims. Marijuana was a perfect scapegoat to explain increasing crime rates. The Federal Bureau of Narcotics claimed that marijuana caused crimes of violence and led to insanity and heroin addiction.

The anti-marijuana movement had strong racial overtones. Mexican and black workers in the southern US often took solace in marijuana smoking, and suffered severely at the hands of the white narcotics police. Even the medical establishment supported the racism. A 1931 issue of the *New Orleans Medical and Surgical Journal* stated that "The dominant race (meaning whites) and most enlightened countries are alcoholic, whilst the races and nations addicted to hemp and opium, some of which once attained great heights of culture and civilization, have deteriorated both mentally and physically." Such absurd statements totally ignored the scientific evidence that was already available at the time.

Thirty years earlier, the British government had established the Indian Hemp Drugs Commission to answer questions about the use of marijuana in India, then under British rule. The expert committee interviewed almost 2,000 witnesses, made field trips to thirty cities, and published a thorough seven-volume report. It concluded that small doses of hemp were beneficial, and that moderate use of cannabis had no significant injurious mental, physical, or moral effect. Furthermore, even abuse of cannabis was less harmful than the abuse of alcohol. The commission recommended a system of licensing and revenue taxation for the sale of cannabis and suggested that overly restrictive marijuana laws would drive people to more dangerous drugs like alcohol and opium.

The Indian Hemp Commission report was a thoroughly researched, levelheaded account of marijuana use. It was totally ignored in the US because it did not fit the political ideology of the times. It was far more suitable for the Federal Bureau of Narcotics to paint a picture of marijuana as a ghastly and dangerous substance in order to push for the establishment of "narcotics farms for the confinement and treatment of persons addicted to Indian Hemp."

The vestiges of that era are still with us. Some right-wing fringe groups attribute the moral decay of our society to marijuana use. Smoking pot damages the brain, they argue. It leads to harder drugs. Pro-marijuana groups have fought back, starting with the beatniks of the 1950s and the hippies of the 1960s. Smoking pot is not only pleasurable and innocuous, they claim, but it also has decided health benefits. It should be legalized.

What do the scientific facts say? Marijuana does not destroy the brain, but heavy, daily use may lead to slight memory impairment. A well-controlled study carried out at Harvard University examined sixty-five heavy users and found that, when compared to light users, they had a slightly harder time with some card-sorting experiments. Marijuana does impair dexterity and visual skills and therefore affects driving, but does not make people drive recklessly, as alcohol does. Nor is there evidence that it leads to the use of harder drugs. In fact, in Holland, where it has been legal to purchase marijuana in coffee shops since 1976, there is an amazingly low rate of heroin addiction. While more Dutch teenagers try marijuana, fewer go on to be regular users than in the US.

Obviously, marijuana smoking is not good for the lungs. These organs were designed to breathe clean air. Studies have shown that three to four joints is roughly equivalent to twenty cigarettes in tar content, mostly because of the lack of filters and the tendency of pot smokers to hold the smoke in their lungs for a longer time. Unlike tobacco cigarettes, however, marijuana does not lead to blocked airways or emphysema. Chronic bronchitis, though, is a possibility. Marijuana has been found to have estrogenic effects, and according to some, it may play a role in the reduction of sperm counts. There is evidence that constant marijuana use leads to a generalized lack of motivation to pursue studies or a career. In some cases, smoking pot can cause anxiety and panic. All of which are good reasons to be extremely wary of marijuana use.

On the other side of the ledger, marijuana has been hailed as an effective way to control the nausea and vomiting associated with chemotherapy. Controlled trials have shown that it is indeed effective. Its main ingredient, delta-9-tetrahydrocannabinol, has been available in a purified drug form, under the name dronabinol, since 1985, but has not seen wide use. Proponents of marijuana smoking claim that inhalation of the smoke is far more effective than taking the pill form. Indeed, they could be right, as the liver tends to metabolize the oral medication quickly. But today, other very effective anti-nausea medications are available.

There is also evidence that smoking marijuana may have some anticonvulsive effects, such as in the treatment of epilepsy. In a celebrated legal case in Toronto, Terry Parker was acquitted of charges of possession of marijuana when he convinced the judge that he was growing the plants for his own use. His epilepsy, he claimed, was better controlled than with the usual prescription drugs. There is also evidence that marijuana can ease the pain of multiple sclerosis, and that it may even have an effect in controlling the spasticity sometimes associated with the disease. Based on controlled trials, Canada has approved an oral spray form of marijuana (Sativex) for the relief of symptoms associated with multiple sclerosis.

Marijuana has also been used for the treatment of glaucoma. Until 1991, some patients were actually given prescriptions for joints to reduce the pressure in their eyes. The availability of better medications has made this approach unnecessary, although there have been cases in which patients responded better to cannabis than to other drugs. The appetite stimulant effect of marijuana may be useful in the wasting often seen in AIDS patients, but pure dronabinol is as effective as smoking. Obviously, marijuana is not a miracle drug. But nor is it a "killer weed."

The World Health Organization actually concluded that marijuana is safer than alcohol or tobacco, but its report was suppressed when officials from the US National Institute of Drug Abuse objected to the findings. This in spite of the fact that both the *New England Journal of Medicine* and *The Lancet*, perhaps the two top medical publications in the world, recently published editorials favoring the liberalization of marijuana laws. Oh, yes, the marijuana issue is still very politicized. Lee Brown, who headed the US Office for National Drug Control Policy under President Clinton, once returned from the Netherlands and offered this remarkable view of the Dutch situation: "I've visited their parks. Their children walk around like zombies." Shades of *Reefer Madness*. It seems that the more things change, the more they stay the same.

OBSESSIONS AND COMPULSIONS

Cats do not climb into washing machines. Nor do they take refuge in dishwashers, ovens, or refrigerators. The young chemistry graduate student knew this full well. Yet he was unable to turn on his washing machine before checking over and over again to make sure one of his pet cats had not wandered inside. He couldn't take food out of his fridge without repeatedly opening and closing the door until he was convinced that a cat had not accidentally become trapped. Of course he knew that this behavior made no sense at all. Nevertheless, he was compelled by some inner force to check and recheck. Such is the bizarre nature of obsessive-compulsive disorder, or OCD.

Can you imagine a twelve year old having to touch a doorknob in his room exactly 375 times before being able to go outside, or a teenager doing sit-ups, unable to stop, sometimes screaming in pain, until he had completed a self-prescribed number? How about the lady who saw a cockroach run across the floor of a supermarket, and for the next fourteen years, washed all her groceries—as well as everything else she brought into the house—in an attempt to protect herself from germs? Or the man who scrubbed his hands as many as fifty times a day, unable to stop, even though they were inflamed and bleeding? What about the teenager who couldn't help counting everything he came across: telephone poles, passengers in passing cars, letters on signs, or words in a sentence? These things are unimaginable—except, of course, to those suffering from OCD. Then they are very real—real enough to destroy lives.

First, a couple of definitions are in order. Obsessions are unwanted, unreasonable, intrusive thoughts that can poison virtually every waking moment. Fear of contamination by germs, fear of accidentally doing harm to others, and a fear that terrible things will happen if something is left undone are typical examples. Compulsions are rituals that are performed in some hapless attempt to gain relief from obsessive thoughts. A sixty-year-old man, for example, suddenly developed an obsession about garbage on the street, and became convinced that if he did not pick up every bit he encountered, some horrific calamity would befall him. He began to collect whatever he could in garbage bags, which he then stored in his house. He eventually lost his job because he was spending so much time picking up garbage that he never made it to work. This gentleman was intelligent, totally aware of the ridiculous nature of his activity, but was powerless to do anything about it. So was the sixty-five year old who spent some twenty-five years trying to sneak a peek at other men's penises. He had no homosexual urges at all, but was obsessed by the thought that his life would be destroyed if he did not meet his daily quota. Sometimes he would have to drive from truck stop to truck stop, waiting in the bathroom, to try to glimpse a view. Relief came only when he met his quota. Then he would have to start all over again the next day.

These people are most assuredly not insane, but they are mentally ill. Often, they are very intelligent, which makes their torment even more difficult to accept. So then what causes roughly 2 percent of the population (more than suffer from Alzheimer's disease) to be afflicted by this terrible condition? One thing is clear: it has nothing to do with any repressed memories, feelings of guilt, or overbearing parents. There are no subconscious mental conflicts involved. Lady Macbeth trying to wash away her guilt is not a model for OCD. No recorded case exists in medical literature of anyone having been "cured" of OCD through psychoanalysis. But there are numerous cases of victims who have been helped by behavioral and pharmaceutical intervention. The consensus now is that OCD is, like other physical ailments, caused by something having gone awry in the body's complex chemistry.

What evidence is there for this? First of all, the condition is sometimes precipitated by physical injury. Blows to the head, epileptic seizure, and strokes have been known to trigger OCD. Swiss researchers report a fascinating case of a political journalist who had no particular interest in food until he suffered a brain hemorrhage. Following his recovery, he began to compulsively think about eating, and even switched careers to become a food columnist. This tweaked the researchers' interest, and they began to assess patients who had presented with various brain lesions for what they now called "Gourmand syndrome," the development of a sudden obsession with food. Over a three-year period, they uncovered thirty-six such patients, thirty-four of whom had lesions in the right anterior area of the brain. Even more interesting evidence comes from PET (positron emission tomography) studies of the brains of OCD patients. This technique measures brain activity and has confirmed that in OCD parts of the prefrontal cortex and parts of the basal ganglia, particularly the pecan-sized "caudate nucleus," are overactive. When patients improve after treatment, this activity is reduced. The fact that medications that specifically increase serotonin levels in the brain help with OCD also suggests a chemical malfunction.

Then there is the case of the twenty-two year old who was so frustrated by his OCD that he attempted suicide. He shot himself in the head, but survived. Amazingly, along with part of his brain, his OCD also disappeared! Obviously, this is not recommended treatment. What, then, can people who suffer from OCD do? I developed an interest in answering this question many, many years ago, when I first started teaching organic chemistry. I'll never forget my first class.

I was young, fresh out of graduate school, and ready to wow the class with a lecture I had spent hours and hours preparing. After a brief introduction, I turned around to write my name and office number on the board. That's when it happened. I heard a loud bark! I hadn't seen a dog in the classroom, so I couldn't imagine where the sound had come from. Looking around revealed nothing unusual. When I turned to face the board again, the barking started up once more. This time it was followed by a loud string of obscenities! Ruling out the possibility of foul-mouthed canines, I scanned the classroom more carefully. My eyes came to rest on a student sitting in the back with his hands clamped over his mouth, in an obvious state of distress. Then the hands suddenly dropped, and he let out a horrific bark, followed by, to my great astonishment, an obscene gesture with his finger. And that was how I was introduced to the intricacies of Tourette's syndrome, one of the most fascinating of all mental diseases.

Up to that fateful day back in 1973, I had never heard of the condition first described by French neurologist Gilles de la Tourette in 1884. But my ignorance didn't last very long. Numerous conversations with my student revealed that he was highly intelligent, but tormented by the uncontrollable urge to grunt and bark (vocal tics), grimace (motor tics), scream socially unacceptable phrases (coprolalia), and make obscene gestures (copropraxia). Something was obviously askew in his brain; some chemistry was going wrong. Once we had explained this curious affliction to the class, his symptoms improved. I suppose that the stress of trying to hide the condition that first day probably brought on the extreme manifestations, and scared me half to death.

Eventually I learned that about half of all Tourette's patients also suffer from obsessive-compulsive disorder (OCD). Unfortunately, my poor student fell into this category. He had difficulty finishing exams because he could not go on to the next question until he was convinced that he had gotten the previous one perfect. Pencils and pens on his desk had to be lined up in order of increasing size, and any disruption of this arrangement would bring on signs of terrible internal turmoil. This really stirred my interest, because I also had a friend who was compulsive about keeping things "just right." I had never considered this to be a disease, and I must even admit to a little mischievous enjoyment found in watching him scramble after I revealed I had rearranged something in his house. Thanks to the episode with my student, though, I now realized that OCD was no laughing matter.

Tourette's and OCD are related in the sense that both are due to some faulty brain chemistry, and both have a genetic component. The chemical faults, however, are not the same, and a person can suffer from either condition without being afflicted by the other. Some of the symptoms of Tourette's are similar to those of schizophrenia, a mental illness partly attributed to high levels of the neurotransmitter dopamine. That's why medications that block dopamine activity were the first ones tested in the treatment of Tourette's syndrome. Haloperidol (Haldol) was adept at reducing the symptoms, but caused Parkinson's-like side effects. Excessive norepinephrine activity has also been associated with tics, and clonidine (Catapres), which controls norepinephrine levels, has proven to be useful in Tourette's, as well as in the treatment of OCD.

Depression is a common side effect of obsessive-compulsive disorder, and its treatment with clomipramine (Anafranil), an antidepressant, resulted in the first truly effective drug against OCD. Physicians noticed OCD symptoms were alleviated in patients treated with clomipramine. Since this drug raises serotonin levels in the blood, the search was on for other medications that would do this even more effectively. The selective serotonin reuptake inhibitors (SSRIs), like fluoxetine (Prozac), fluvoxamine (Luvox), sertraline (Zoloft), and paroxetine (Paxil), fit the bill, and have become the therapy of choice for OCD.

In rare cases, in children, OCD can be triggered by a strepto-coccal infection. White blood cells produce antibodies to fight off such an infection, but they may sometimes mistakenly attack and destroy cells in the caudate nucleus, an area of the brain known to function improperly in OCD patients. In such cases, passing the blood through a machine that filters out the rogue antibodies can afford relief from OCD, as can treatment with antibiotics. The best hope, though, to help people tortured by OCD lies in behavioral therapy, whereby they have to confront their fears. A person terrified of germs is encouraged to handle some object—money, for example—that obviously harbors them. They are then asked to resist the urge to wash their hands immediately. At first, it's a real struggle, but eventually, by increasing the time until they give in to the compulsion to wash, they learn that nothing terrible happens. Usually a combination of behavioral and drug therapy makes OCD manageable, but a total "cure" is rarely achieved. Patients commonly see several doctors before a proper diagnosis is made, and take an average of seventeen years from the onset of symptoms to find appropriate treatment.

My student was eventually treated with haloperidol and clomipramine and went on to a successful career as an engineer, although he still has the occasional urge to bark during meet-ings. But, having learned from the experience in my class, he now informs his colleagues of his condition so that they are not shocked by the sounds or obscenities.

WATER BOTTLE CONFUSION

My readers and listeners pose all sorts of interesting questions. Like whether or not it is true that water bottlers are putting a risky hormone called diethylhydroxylamine (DEHA) into the plastic used to make the bottles so that people will feel younger and buy more of their product. When I heard this question, I knew that it was nonsense, but it took a fair bit of detective work to uncover how this remarkable query arose. Let me share it with you.

I'm quite sure the starting point is a scare circulating on the Internet about the migration of a chemical commonly referred to as DEHA into water from plastic water bottles. The essence of the message is as follows: "the plastic used to make these bottles contains a potentially carcinogenic element (something called diethylhydroxylamine or DEHA), which leaches out of the plastic on repeated washing and rinsing." Consumers are warned that such water bottles should not be refilled but instead be discarded after a single use.

Right off the bat, there are several issues here. First of all, DEHA, it is not an "element." It is a compound. And it certainly is not a hormone. And in any case, the writer of this epic epistle has the wrong compound. Diethylhydroxylamine is indeed sometimes abbreviated as DEHA, but it has nothing to do with plastic water bottles. The chemical in question is diethyl-hexyladipate, commonly, and perhaps confusingly, also abbreviated as DEHA. This is an approved plasticizer, a substance added to certain plastics to make them soft and pliable. But neither of the "DEHAs" is classified as a carcinogen. Then there is a further problem with the scare. While diethylhexyladipate is commonly used as an additive in certain plastics, it is not an ingredient in the polyester used to make water bottles! Polyester is innately flexible and does not require plasticizers.

So how did a compound that isn't even present in plastic bottles trigger the alarm? As with many other such Internet scares, there is a kernel of truth that has been blown completely out of proportion. In this case, that kernel is found in an abstract of a talk given by a master's student from the University of

Idaho at a scientific conference. Such abstracts are not subject to any peer review and are not considered to be a form of scientific publication. The intent is that the information presented will eventually be submitted as a paper to a journal, where it will undergo appropriate review by experts in that field of research. As far as I can tell, that never happened in this case, probably because the study performed did not have the scientific rigor required for publication. The student investigated contaminants in bottled water and found a number of organic compounds, including diethylhexyladipate. Apparently unaware that this is not used in polyester bottles, the author assumed it was leaching out of the plastic. Stimulated by this, the Idaho student examined a variety of other plastic bottles and found DEHA in the water they contained, leading to the conclusion that "migration of DEHA was not limited to polyester bottles alone and other bottles may also pose a health hazard."

But a critical control experiment was never performed. Was there any DEHA in water stored in glass bottles or in water that came from the tap? Actually, there is. Because DEHA is a ubiquitous plasticizer used in the manufacture of items ranging from toys to shower curtains, it shows up in trace amounts everywhere. You can find it in food, clothing, and water. We know this because the Swiss Federal Laboratories for Materials Testing and Research studied this issue extensively. All sorts of water samples that had never been in contact with any plastic bottle showed traces of DEHA! Whether the samples came from plastic bottles or glass bottles, they contained the inconsequential amount of about 0.01 to 0.05 parts per billion. The World Health Organization has set a maximum of 80 ppb for DEHA in drinking water, so plastic bottles are simply not an issue here. At least they're not as far as DEHA goes. There may be reasons not to refill water bottles, but that has to do with possible bacterial contamination, not with the leaching of diethylhexyladipate.

So far, so good. But what about the business of DEHA making people feel younger? I puzzled over that one for a while. I think that someone's fingers dancing on the keyboard while they were "researching" the DEHA–water bottle connection made a little slip and typed DHEA instead of DEHA. All of a sudden, claims of the potential rejuvenating properties of dehydroepiandrosterone (DHEA) began to frolic on the screen. A "fountain of youth," many websites claim, a "superhormone!" And what is this miracle? DHEA is a naturally occurring compound synthesized from cholesterol in the adrenal glands. Production peaks in the twenties, and then declines so that by the eighth decade of a person's life, the amount of circulating DHEA is only 20 percent of that found during the vigor of youth. (Now don't go jumping to the conclusion that popping cholesterol will increase DHEA in the blood and thereby make you live longer.)

Reasonably, though, researchers began to explore the possibility that maintaining the DHEA in the blood at levels found in young people may help avert some of the problems of aging. Animal studies showed some intriguing results in terms of delaying cancer and the hardening of arteries. A widely reported study in humans showed that taking 50 milligrams of DHEA for three months resulted in an improved feeling of well-being. But there is another side to this coin. Researchers worry that DHEA, a relative of testosterone, may increase the risk of prostate cancer in men and cause facial hair growth in women. DHEA is illegal in Canada but can be freely sold as a "dietary supplement" in the US. Curiously, people who would not consider taking prescription hormone replacement therapy seem to uncritically jump on the DHEA bandwagon.

Now, to get back to answering our original question: DHEA is not the same as DEHA, and neither substance is present in the polyester used to make water bottles. And I really don't think that manufacturers are sneaking rejuvenating hormones into the plastic in order to increase sales. Such silly, unreliable "they say" stories do, however, cause me to grow a few gray hairs. Hmmm . . . maybe I should try drinking more bottled water.

A Toxicological Nuance

Be warned! The food, cosmetic, and pharmaceutical industries commonly use an industrial solvent that remains in the final product. Used as a fire retardant, stain remover, and antifreeze ingredient, it cannot be removed from fruits or vegetables by washing. It is always found in malignant tumors, and is responsible for thousands of deaths every year through inhalation. Surveys show that people become quite indignant when informed of the widespread use of this chemical, and are quite willing to sign petitions to have it banned. But banning dihydrogen monoxide would certainly not improve our life. Water is hardly a poison! Yet everything I said is true, including the thousands of deaths every year through inhalation, perhaps better described as "drowning."

It is obviously easy to conjure up scary scenarios that will alarm people by using scientific lingo selectively and inappropriately. Numerous publications and Web sites do this as they warn us about hazardous chemicals found in our foods, cosmetics, and cleaning agents. We're told that parabens, commonly used preservatives in cosmetics, are "estrogen mimics," that polyethylene glycol, used as a thickener in shaving cream and as an emollient in skin lotions, may be contaminated by 1,4-dioxane, a known carcinogen, and that all perfumes contain toluene, which can cause liver, kidney, and brain damage.

All of these statements are technically correct, but their practical relevance is highly suspect. Remember that the pivotal credo of toxicology, first voiced by Paracelsus in the sixteenth century, is that "only the dose makes the poison!" Yes, perfumes do contain toluene, but in amounts that are way below levels that cause any effect. Evidence for the presence of a substance is not evidence of harm. After all, we don't avoid apples even though their seeds harbor the deadly toxin cyanide, we happily eat strawberries although they contain acetone, a known neurotoxin, and we are not deterred from toast by the presence of 3,4-benzopyrene, an established carcinogen. The toxic properties of these chemicals are indeed real. When test animals are exposed to high doses of acetone or 1,4-dioxane, they certainly do show neurological damage and tumor growth. But that doesn't mean small doses in humans over a longer time will have a similar effect. In fact, they may have a significantly different effect.

Buckle your seatbelts, because we are going on a bumpy ride. We are going to rattle some of the basic tenets of toxicology. Not only may tiny doses of toxins not be dangerous, they may actually be good for us! Admittedly, that sounds outrageous. So let's set the stage for an exploration of a revolutionary concept known as "hormesis," the notion that small doses of toxins can be healthful. Actually, this may not be quite as surprising as it first appears. Vitamin D is certainly healthy in small doses, but large doses can be lethal. A glass of red wine a day may prolong health, but a bottle a day increases the risk of various cancers. Is it possible, then, that exposure to tiny doses of really toxic substances, such as dioxins or pesticides, may actually lead to a better-functioning immune system? Ed Calabrese, professor of toxicology at the University of Massachusetts, thinks this is a real possibility. And he is no academic slouch. Dr. Calabrese has published over 500 research papers and is recognized as the world authority on exposure to trace chemicals.

Calabrese originally got interested in low-dose toxins when, as an undergraduate, he became involved in a project to investigate the amount of herbicide needed to stunt the growth of peppermint plants. Much to his surprise, the plants grew taller when sprayed with the chemical. It turned out that the solution had been improperly prepared and was far more dilute than intended. Years later, at a conference on radiation, Calabrese was reminded of his peppermint plants when he learned that studies had shown people exposed to low-level radiation lived longer and had lower cancer rates. He then began to scour the scientific literature for other such effects and came up with almost 6,000 publications that had documented dramatically different effects at very low concentrations of toxins when compared with those seen at higher doses. Rats exposed to traces of the insecticide DDT, or the treacherous pollutant dioxin, developed fewer liver tumors than unexposed rats. Bacteria frolicked when treated with trace amounts of antibiotics.

Hormesis (from the Greek "to excite") is still controversial, because if such effects are real, we may have to reevaluate our exposure standards for contamination of air, water, food, and soil by certain chemicals. It does, however, make biological sense. When attacked by poisons, an organism responds by unleashing a variety of molecules, mostly enzymes, which attempt to repair the damage. If the amount of toxin is minute, there may be an overreaction, with more defense chemicals being churned out than are needed, leaving an excess to deal with the molecular insults of everyday life. It may yet turn out that the apocalyptics who warn us of the perils of exposure to parts per trillion of "toxic" chemicals are on the wrong track. Of course, nobody is suggesting doping our bodies with traces of DDT or dioxin, but accumulating information suggests that the paranoia about trace amounts of toxins in our environment is unjustified.

And just to show you how far irrational chemophobia can spread, the city of Aliso Viejo, California, nearly banned Styrofoam cups when a paralegal working for the City Council chanced upon a "professional looking" Web site that listed the evils of dihydrogen monoxide and described its use in the production of Styrofoam. The council was ready to ban Styrofoam—that is, until someone pointed out that the evil chemical was just water.

SOME CROOKED CHEMISTRY

One of my favorite detective stories is Agatha Christie's *The Crooked House*. The plot revolves around an elderly tycoon who requires daily insulin shots. He also suffers from glaucoma, for which he has been prescribed eyedrops. Everything is fine until someone in the "Crooked House" switches the eyedrops with the insulin. Murder most foul!

Does the chemistry make sense? Let's do a little detective work of our own. As clearly stated in the novel, the eyedrops contain physostigmine, a substance introduced in the late 1800s for the treatment of excess pressure in the eyeball, a symptom characteristic of glaucoma. Physostigmine, or "eserine," as it is also known, opens up the tiny ducts through which excess fluid is normally expelled from the eye. Could this drug really be lethal if injected into the bloodstream?

Physostigmine has a long and interesting history. It is the active ingredient in the "ordeal" bean, found in the Calabar region of Nigeria. Why the term "ordeal" bean? Because it was traditionally used by certain tribes as a test of guilt. Someone suspected of having committed a crime was forced to swallow a handful of beans. If he died, he was guilty. Unfortunately, he probably died even if he wasn't guilty. Physostigmine is known to enhance the activity of acetylcholine, a chemical essential for the proper functioning of our nervous system. It does this by inactivating an enzyme called cholinesterase, which normally degrades acetylcholine after it has done its job. The result is a buildup of acetylcholine, which can lead to paralysis of the respiratory muscles and death. Maybe if the accused were really confident of his innocence, he would eat the beans quickly, vomit, and survive!

By the late 1800s, physostigmine had been isolated from the Calabar bean and was widely used in the form of eyedrops for the treatment of glaucoma. The amount needed on a daily basis was very little, but there certainly would have been enough of the active ingredient in a bottle to kill if directly injected with a syringe. Had a physician arrived soon enough, the effects probably could have been reversed. Atropine, found in the belladonna plant, can block the receptor sites on nerve cells that are normally activated by acetylcholine. This antidote was routinely carried by the doctors of the day, not necessarily to deal with physostigmine poisoning, but because atropine is a potent heart stimulant. But if the victim had lived, there would have been no murder, and no story. Who needs a story, though, when real-life poisonings may be stranger than fiction!

Just consider the celebrated murder trial in which the prosecution began its case with an unusual demonstration, one that would never be condoned today. But back in 1893, the judge agreed for a cat to be brought into the courtroom, and for it to be given a lethal dose of morphine. As its life ebbed away, a few drops of belladonna extract were applied to its eyes. Why was such a cruel experiment necessary? Because the prosecutor wanted to demonstrate to the jury that the characteristic pinpointing of the pupils of the eye caused by morphine could be reversed with atropine.

Dr. Robert Buchanan had been accused of murdering his wife after having struck up a relationship with the wealthy proprietress of a brothel. In 1892, Mrs. Buchanan developed a sudden illness and died. The medical examiner concluded that the cause had been brain hemorrhage, and refused to consider the possibility of murder, even though one of the brothel owner's former lovers insisted that the doctor had done away with his wife, probably with morphine. But the medical examiner maintained this was impossible because the victim had shown no signs of pinpoint pupils. When the *New York World* newspaper began to question the treatment of the case, the coroner was forced to order an exhumation. The remains were found to contain enough morphine to have caused death, and Buchanan was put on trial for murder. His undoing was a conversation that was reported by a witness. The doctor, it seems, had railed against another accused morphine poisoner as an incompetent. His downfall had been the telltale pinpoint pupils of the victim's eyes. The fool, Buchanan had said, should have applied some belladonna! When this story came out, another witness recalled having seen Dr. Buchanan put some sort of drops into his wife's eyes before she died. The good doctor was convicted and executed in the same electric chair in which the "bungler" had met his end two years earlier.

Now let's fast-forward about 100 years and get acquainted with Dr. Paul Agutter, who probably would have appreciated Buchanan's efforts. A well-known biochemistry professor in Edinburgh, Scotland, he, like Buchanan, tried to do away with his wife in a chemical fashion. He figured a high dose of atropine would do the job, and put some into the tonic water she always had with her gin. This was a good choice, because the bitter tonic disguised the taste of the atropine. To throw the police off his trail, he also adulterated some bottles of tonic water in the local supermarket with small doses of atropine to suggest that a mass poisoner was on the loose. But Dr. Paul Agutter was unlucky. One of these bottles happened to be purchased by the wife of an anesthetist. When she and her son became ill, he recognized the symptoms as atropine poisoning! Within the next few days, five other locals were diagnosed with the same condition. All survived after they were appropriately treated. The police tested their tonic water and found atropine, but not as much as was in the bottle belonging to Mrs. Agutter. Paul Agutter finally confessed to attempted murder and was sentenced to twelve years in prison. Why did he want to murder his wife? It was the age-old reason. He wanted to inherit her family fortune and then marry his mistress. He probably would have gotten away with it, too, had one of the bottles not ended up in the household of an alert anesthetist.

THE CRYONIC QUEST FOR IMMORTALITY

A television image from the 1960 baseball season is indelibly etched in my mind. It features Ted Williams striding to the plate for the last time and slamming a home run into the right-field bleachers. Fittingly, the "Splendid Splinter" left the game in a blaze of glory. Unfortunately, he left the game of life in a blaze of controversy. Today, Williams' fans have to contend with a different image of their hero. That of "Teddy Ballgame's" body and head resting in separate liquid nitrogen–cooled containers at the Alcor Life Extension Foundation in Scottsdale, Arizona.

Alcor is a "cryonics" company, meaning that it stores bodies and heads at very, very low temperatures, hoping to preserve them until scientists of the future find a way to "reanimate" them. It's a long shot, the cryonicists admit, but at least it's a shot. The alternatives, they suggest, are not particularly attractive: you could either be incinerated into dust, or served up as a meal for bacteria and worms.

The quest for immortality is nothing new. Ancient Egyptians thought mummification was the answer, the alchemists believed the secret lay in finding a way to make gold, a metal they considered to be immortal, and Ponce de Leon searched in vain for the fountain of youth. But the introduction of any vestige of science into this quest came with the development of methods to reach very low temperatures. At -196°C (-321° F), which is the boiling point of liquid nitrogen, essentially all biological activity stops. In theory, bodies or body parts can be kept from ever decaying. "Reanimation," though, is quite another matter.

Back in 1965, physics teacher Robert Ettinger got the ball rolling with his book *The Prospect of Immortality,* in which he laid out his theory about cryonics. By 1967, a company had been founded, and the first patient, if that's the right word, had been frozen rock hard. James Bedford, a psychology professor from California, now awaits reanimation, along with some ninety others who have invested anywhere from $40,000 to $130,000 for a chance to be thawed out in the future to resume their lives. For some reason, the company insists on receiving payment up front. I suspect, though, that the reanimated people will not have any financial problems if they wake up. Writing books, granting interviews, and appearing on talk shows (if those still exist) should be pretty lucrative. People will certainly be interested in hearing about our primitive way of life.

Preserving the body, according to the cryonicists, is not really necessary. Only the head is needed (and that can be frozen for the bargain price of $40,000). If indeed future scientists find a way to bring frozen tissues back to life, they will certainly have also found ways to clone the body from cells. But the brain is unique. This is where our memories, hopes, fears, and bizarre thoughts about immortality lie. That's why a number of the patients are just "neuros," the term used for heads with no bodies.

It comes as no great surprise that biologists and cryonicists don't see eye to eye. When water in tissues freezes, it expands. This can cause blood vessels to burst. And as water outside cells freezes, the concentration of solutes in the remaining fluid increases. This then causes water to flow out of the cells by osmosis, resulting in the collapse of the cell. Alcor representatives maintain that they have solved such problems by developing a technique through which a "cryoprotectant" (essentially antifreeze) solution is injected immediately after death and before cooling. This, they claim, results not in traditional freezing, but in "vitrification," without the formation of ice crystals.

To date, there is not much evidence that Alcor will be able to deliver the goods. We can't even preserve kidneys, livers, and hearts destined for transplants for more than a few days. And they certainly can't be frozen. Sperm cells, on the other hand, can; farmers commonly use frozen sperm for artificial insemination, and human sperm remains viable after freezing. The very first step in the Alcor quest, it seems, should be to freeze a live mammal and then thaw it out to see if it can be brought back to life. Nothing like that has been done. Obviously, there are ethical issues involved with such experiments as well. And Alcor has had its experiences with that. Back in 1987, the company was accused of having removed Dora Kent's head before she was dead. Dora was the mother of Saul Kent, one of the earliest proponents of cryonics. He had her transported to the Alcor facility, then in California, so that she could be decapitated and processed immediately after death. The Coroner's Office labeled the death a homicide after barbiturates were found in the headless body. The head itself was never located, and eventually charges were dropped.

An even more interesting case is that of Thomas Donaldson, a computer consultant who was diagnosed with an inoperable brain tumor in 1988. Donaldson decided that his only chance was to have his head removed immediately, before his whole brain was ravaged by cancer. Realizing that the technicians involved could be accused of murder, Donaldson petitioned the state to allow the process by which he would be anesthetized and placed on a heart-lung machine as his blood was replaced with a chemical solution. His head would then be removed and frozen. The State of California denied the petition and a subsequent appeal. As it turned out, it's a good thing Donaldson kept his head. His tumor went into remission, and he is still alive. And how will Ted Williams, the last man to bat over .400 for an entire season, eventually fare in the hands of Alcor? If I were a betting man, I would bet on a lifetime average of .000 for the cryonicists.

The Cold Facts about Antifreeze!

Oh, what a difference a single carbon atom makes! It can be the difference between life and death! Just 2 tablespoons of ethylene glycol can kill a child, while the same amount of propylene glycol is practically harmless. These compounds are not esoteric substances that are only of theoretical interest; they are commonly encountered in various antifreeze preparations. Ethylene glycol is found in virtually every automobile radiator and is responsible for a couple of dozen deaths every year in North America, along with thousands of cases of poisoning. Since it can cause inebriation just like ethanol, some desperate alcoholics resort to ethylene glycol with the mistaken belief that it is no more toxic than the usual alcohol they drink. But it is! Other poisoning cases involve suicides or the accidental ingestion of the sweet-tasting glycol by children. There are also innumerable cases of poisonings of pets that love to lap up little puddles of the fluid that leak from a car's radiator.

Antifreeze formulated with propylene glycol (Sierra brand) is available, albeit more expensive than the conventional variety. Many zoos have converted to using propylene glycol in the radiators of their vehicles to reduce the risk to animals that might ingest the fluid after radiator leaks or boil overs. The switch to the safer antifreeze received a big boost when a rare California condor died after drinking from an ethylene glycol spill. So how could just one little extra carbon atom in a molecule cause such a difference in toxicity? It all comes down to the manner in which ethylene glycol and propylene glycol are metabolized in the body.

The liver is the body's main detoxicating organ, and it is here that an enzyme, alcohol dehydrogenase, reacts with glycol of either kind. But the products of the reaction are quite different! Propylene glycol is converted into lactic acid, while ethylene glycol mainly forms glycolic and oxalic acids. Any excess acid in the blood is a problem, but oxalic acid presents a further complication. It reacts with calcium ions in the blood to form microscopic crystals of calcium oxalate, which can interfere with blood flow and cause damage to the brain, heart, and lungs. But the most serious effect is blockage of the tiny tubes that make up the blood-filtering structure of the kidneys. Death from ethylene glycol poisoning is usually due to kidney failure.

If ethylene glycol poisoning is diagnosed quickly, it can be treated effectively. A stomach wash with saline solution and activated carbon (a form of charcoal that absorbs toxins) can remove some of the glycol before it is absorbed. Intravenous sodium bicarbonate can counter the metabolic acidosis. And then comes the most interesting part of the treatment. Oral and

intravenous administration of alcohol! Since alcohol dehydrogenase has a greater affinity for alcohol than for ethylene glycol, the enzyme goes to work on alcohol and leaves the glycol alone. By the time the alcohol has been metabolized, the ethylene glycol has hopefully been eliminated in the urine. The patient ends up very drunk, but very much alive! A drug called fomepizole (4-methylpyrazole) can also be used to inactivate alcohol dehydrogenase without the adverse effects of alcohol administration.

With this background we are now ready to tackle the "Swiffer WetJet" scare story that has been making the rounds on the Internet. This mop-like device uses a premoistened cloth to clean stains on floors. A circulating e-mail claims that the solvent used on the cloth is dangerous to pets, and tells of a dog and two cats that perished just from having walked on a freshly mopped floor.

I'm going to come to the aid of Procter & Gamble, the makers of the Swiffer WetJet. Not because I'm a great supporter of giant corporations, but because I'm a great supporter of good science. And the allegations against the Swiffer WetJet are just plain bad science. The e-mail does not document where the supposed death of the dog occurred, so it is impossible to check out the story or the claim that an autopsy was performed on the dead animal and liver failure was diagnosed. In any case, according to the author, the toxic culprit is "propylene glycol n-propyl ether," the solvent used in the Swiffer product. This, he suggests, is "one molecule away from antifreeze," which is known to be deadly to animals.

First of all, the antifreeze that really is dangerous to dogs, as we have seen, is ethylene glycol, but it causes kidney and not liver failure. That, though, is hardly the point. The statement of propylene glycol n-propyl ether being one molecule away from antifreeze is absurd. This is where a little learning becomes a truly dangerous thing. Propylene glycol is close in structure to ethylene glycol. It isn't "one molecule away," it just has an extra carbon and three hydrogens. That difference, though, is what makes it into a safer product! That's why propylene glycol has replaced ethylene glycol in many products where toxicity is an issue.

But we are still not done. Swiffer doesn't contain propylene glycol; it contains propylene glycol n-propyl ether, a completely different substance with very low toxicity. Furthermore, the solvent on the mop is mostly water with the propylene glycol n-propyl ether being present to a maximum of 4 percent. And finally, the solvent is reabsorbed into the mop, so essentially nothing is left on the floor. The only thing left after using a Swiffer WetJet is the Internet nonsense that is going around. I wish we could swiff that away as easily as the Swiffer WetJet swiffs away stains on floors.

Would You Like to See My Etchings?

The pain in his finger was excruciating. He had never felt any-thing like this, the professional painter told the emergency room physician. Yet the finger, save for a little white discolora-tion, looked almost normal. Perhaps there was a small blood clot blocking the circulation, the doctor thought, and started the patient on aspirin. It didn't do the job. The man was back the next day, his face contorted with pain. But now the finger no longer looked normal; it was red and swollen, and the tip had turned black! It was obviously time for some probing questions. At first, the patient could not think of anything he had done differently, but then he recalled having used a new, heavy-duty cleaning agent a couple of days earlier. In fact, he still had some of it left. A quick perusal of the bottle's label, and the mystery was solved. The active ingredient was hydro-fluoric acid (hydrogen fluoride dissolved in water), one of the most dangerous chemicals you can ever encounter.

Once the culprit was identified, the course of treatment was clear. The application of a 2.5 percent calcium gluconate gel brought instant relief. After three days of such treatment, the finger was as good as new. A lucky man! The fluoride ion released by hydrofluoric acid is terribly corrosive. It penetrates tissues readily and combines with calcium to form insoluble calcium fluoride. Nerve function relies on a balance of calcium and potassium ions, and when calcium is removed, normal con-duction fails, and terrible pain ensues. Calcium gluconate counters the problem by quickly replenishing calcium. If a topical application doesn't help, it can be administered intrave-nously. Prompt diagnosis of HF exposure is critical because it can cause far more problems than severe pain. Sudden calcium depletion can cause changes in heart rhythm, and can lead to death. One of the most frightening features of hydrogen fluoride exposure is that the skin can look practically normal while the underlying tissues are being eaten away.

Not everyone who has a run-in with hydrofluoric acid is as lucky as our painter. A lab technician who spilled the contents of a small bottle in his lap paid the ultimate price. In spite of immediately hosing himself down and jumping into a swimming pool, he sustained chemical burns on about 10 percent of his body. By the time he arrived in hospital, his blood calcium was dangerously low, and he soon lost consciousness. Circulation in his right leg was lost, and it had to be amputated. Despite valiant efforts, he died of hydrogen fluoride poisoning within two weeks. Had he been wearing proper polyvinyl chloride (PVC) protective apparel, he would still be with us today. But this lab technician most assuredly was not the first to succumb to hydrogen fluoride exposure. Indeed, the years that followed the discovery of HF were filled with such tragedies.

Way back in 1771, the Swedish chemist Karl Wilhelm Scheele was investigating the properties of a mineral known as fluor-spar (calcium fluoride). He mixed the pulverized rock with sulfuric acid and heated the mixture. The results were literally staggering! Scheele practically choked on the vapors that were produced, but he survived to make a dramatic discovery. The clear glass vessel in which he had combined the fluorspar and sulfuric acid had become cloudy. Somehow, the glass had partially dissolved! Fluorine forms extremely strong bonds with silicon and strips silicon atoms right out of the silicon dioxide framework of glass. No one before had ever encountered a chemical that could dissolve glass, and now Scheele could produce such a substance on demand. What it could be kept in was the first obvious question. Bottles made of gutta percha, a form of naturally occurring rubber, solved the problem. But another question remained to be answered. Why would anyone want to dissolve glass? Perhaps it was to attract ladies.

"Would you like to come up and see my etchings?" may well be the oldest of all bad pickup lines. Its origins can be traced back to the early use of hydrofluoric acid to etch designs on glass. One shudders to think of all the miseries undoubtedly caused by the cavalier use of such a dangerous substance. But it was not only men who wanted to appeal to the ladies who were interested in etching glass; men who wanted to hide from the gentle sex also saw the appeal of hydrogen fluoride. The Victorian "public house" was the traditional male refuge where drunkenness and raunchy behavior were the order of the day. Clear windows allowed wives and other passersby to witness the debauchery, much to the concern of the revelers. Frosted glass eliminated transparency, and its light-scattering properties made for a very pleasant illumination.

Later, this notion also appealed to lightbulb manufacturers. Clear bulbs produced a great deal of annoying glare. Frosting the bulb with hydrofluoric acid cut down the glare, but the rough surface readily gathered dirt and diminished the light's intensity. So why not put the frosting on the inside? That was tried, but it weakened the glass. Finally, in 1925, Marvin Pipkin discovered that treating the glass with hydrogen fluoride twice instead of only once paradoxically allowed it to keep its strength. We have been basking in the frosted glow of lightbulbs ever since.

In addition to the glass industry, aluminum, steel, and petroleum manufacturers also use large amounts of HF. They take extensive precautions. This is not always the case with consumers, who really would be better off using other rust removers such as phosphoric acid. Not many realize the terrible dangers of hydrogen fluoride–based rust removers and aluminum cleaners. If they did, they would never dispose of a half-empty bottle of the stuff in a garbage bag. That is just what happened in New York. When sanitation workers threw the bag into the compactor on the truck, they got sprayed with hydrofluoric acid. Despite immediate attention, one of the men died within five hours. Manslaughter charges await the perpetrator, should he ever be caught.

Does She or Doesn't She?

"Does she or doesn't she?" wasn't exactly a novel phrase. Many a teenage boy had pondered this question, but it was an advertising agency that turned it into one of the most successful slogans in history. "Only her hairdresser knows for sure" was the ballyhooed answer. The year was 1956, and the reference was to Clairol's introduction of a hair "colorant" that could be applied at home in a single step. Women would no longer "dye" their hair, the ad agency decided, they would "tint" or "lighten" it. "Never say dye" became the agency's motto. Hair "dyeing," you see, was associated with women who lived fast and loose, an image that was not conducive to selling large volumes of product. Tinting, though, was a different story. Within a few years, the percentage of women who colored their hair increased from 7 percent to about 40 percent.

Clairol did not invent the idea of coloring hair. Early Egyptians already used an extract of the henna plant to impart red or orange highlights, and the Romans made a black hair dye by boiling walnut shells and leeks. But it wasn't until chemists learned how to synthesize novel compounds from coal tar that truly effective dyes were developed. Eugene Schueller, a French chemist, gets the credit for creating the first commercial hair dye. Back in 1909, he came up with a basic formulation, very similar to the one used today, that combined para-phenylene-diamine (PPD), ammonia, and hydrogen peroxide. He then founded the French Harmless Dye Company, which a year later was more attractively renamed as L'Oréal. Schueller included "Harmless" in the name because the coal tar dyes that had been used for fifty or so years to color fabrics had already developed a reputation for toxicity. In truth, he had no evidence at all that his concoction was harmless, and claims that PPD and its chemical relatives can have an adverse effect on health have plagued the hair dye industry ever since.

The dye that Schueller created can be referred to as a "permanent dye" because it survives numerous shampooings. Today's permanent dyes are certainly superior to Schueller's, but the basic chemistry is the same. The outside layer of a hair shaft, known as the cuticle, is made of a network of overlapping cells that can be likened to a Venetian blind. For chemicals to seep into the underlying layer, the cortex, where the hair pigments are found, the "blind" must be opened. This is where ammonia comes in. It swells the hair and opens the cuticle so that hydrogen peroxide and the other dye components can get to the cortex. Here, the peroxide gets down to work. Its first role is to disrupt some of the chemical bonds found in eumelanin and phaeomelanin, the natural pigments responsible for black to brown and red to yellowish hair, respectively. These molecules have a variety of carbon-carbon double bonds that can absorb certain wavelengths of light, and therefore determine the color of hair. Hydrogen peroxide reacts with these double bonds so that the altered pigments then reflect most wavelengths of light, and the hair appears much lighter. "Peroxide blonds" like Marilyn Monroe owed some of their fame to the marvels of hydrogen peroxide.

But in a permanent dye, this destruction of melanin is just step one. As the cuticle opens up, molecules that are the building blocks of the eventual dye diffuse into the cortex. There are two distinct species. One, referred to as the "primary," is exemplified by the phenylenediamines, and the other, often an aminophenol, is known as the "coupler." These compounds are stable and do not react with each other until they are mixed with hydrogen peroxide. Once inside the cortex, reaction occurs, and primary and coupler join to form a colored molecule, which is now too large to escape through the slats of the "Venetian blind." It is permanently locked into the hair! The exact color depends on the specific primary and the coupler used. Primaries are usually p-phenylenediamines or p-aminophenols, while couplers are re-sorcinols, m-aminophenols, m-phenylenediamines, or napthols. Want a nice shade of blue hair? Then you'll need to couple m-phenylenediamine with p-phenylenediamine. A combo of resorcinol and p-aminophenol is what you need if you want to find out if blonds do indeed have more fun.

Unfortunately, repeated dying damages the cuticle, leading to roughness and easy breakage, but modern dyes contain conditioners that help to maintain the cuticle's integrity. Thickeners can also be added to ensure that the dye does not run down the face, and ultraviolet light absorbers keep the newly developed color from fading in the sunlight. A further problem with the p-phenylenediamines is sensitization. Some users develop dermatitis on the upper eyelids or the rims of their ears, but in rare cases there may be a whole body reaction characterized by reddening and swelling of the skin. Many European manufacturers have replaced p-phenylenediamine by toluene-2,5-diamine sulfate, which is less of a sensitizer.

Although "permanent" dyes are by far and away the most popular, "semi-permanent" and "temporary" dyes are also available. The semi-permanent ones contain no ammonia or peroxide and are composed of small colored molecules that can diffuse into the cortex. They resist several shampooings. Nitro-phenylenediamines are the most versatile temporary dyes, and although they are similar to compounds found in permanent dyes, they are used in smaller concentrations. Furthermore, some researchers believe that the potential carcinogens in permanent dyes are actually created through the oxidation process. 4-Aminobiphenyl, a recognized bladder carcinogen, can form as an undesired contaminant in permanent dyes. Temporary dyes are composed of molecules that are too large to penetrate the cuticle. Instead, they just stick to the surface of the hair shaft, and can be readily washed away. These dyes are less appealing, in terms of efficacy, but are also less controversial. Aside from the rare possibility of an allergic reaction, they have been shown to be remarkably safe.

The permanent dyes, however, are shrouded in controversy. Some of their component molecules cause cancer in test animals, and several human epidemiological studies have raised the specter of a link to bladder cancer and non-Hodgkin's lymphoma in humans. A highly publicized study at the University of Southern California suggested that women who used permanent dyes regularly over fifteen years tripled their risk of bladder cancer. Sounds ominous, but since the risk of bladder cancer is only about one in 14,000, even a tripled risk is very small. A number of other studies have found no link to any cancer at all. Particularly noteworthy is the Nurses' Health Study, which followed over 90,000 nurses and found no evidence for any association between hair dyes and cancer.

In a paper published in 2005 in the *Journal of the American Medical Association,* McGill University epidemiologist Dr. Mahyar Etminan and colleagues pooled data from seventy-nine studies on permanent hair dyes and found the results comforting. There was no association with any form of cancer. It should also be noted that hair dyes have undergone a number of changes in recent years, and the compounds used now are not the same as the ones that most people used in the studies that suggested a cancer link. Anyone truly concerned, though, can switch to semi-permanent dyes or temporary ones that have not been implicated in the cancer controversy.

Eventually, biotechnology may put an end to this controversy. Wouldn't it be great if a gene that codes for hair coloring could be inserted into hair follicles, the tiny organelles in our scalp from which hair grows? This may not be as outlandish as it sounds! AntiCancer, a California biotech company, has some intriguing preliminary results along these lines. Sure, the results are with mice and not with humans. And the hair is green, but only under blue light. But it's a start. The California researchers were successful in isolating a gene from jellyfish that codes for the production of a protein that glows green in blue light. They then incorporated this gene into an adenovirus and placed a piece of cultured mouse skin into the virus solution. Within hours, a green pigment could be seen in the follicles. When the skin was transplanted to live mice, about 80 percent of the hairs that grew were green! This idea may appeal to some of today's teenagers who favor the idea of fluorescent hair. But there are simpler ways to do this. Like with Kool-Aid. Inventive teens have discovered that this beverage can do more than quench thirst. They've taken to immersing their hair in a hot solution of the stuff to achieve some amazing effects. Of course, one must also take care not to end up with a colored forehead. Cherry flavor is apparently the preferred variety. As one somewhat scientifically challenged teen put it, "I don't like to put chemicals in my hair; I just prefer to use Kool-Aid." Well, he's probably right about the safety business. Kool-Aid is probably more dangerous when drunk than when applied to the hair.

While Kool-Aid may be fine for teens seeking a punk look, it's hardly suitable for the older set interested in covering up the gray. This is where lead acetate comes into the picture. Dyes that promise to banish the gray hairs from your head so gradually that no one will be the wiser actually cover the gray with lead sulfide. The brown-black color of this compound does the trick. The active reagents in these hair products are lead acetate and elemental sulfur. Lead acetate is a water-soluble compound, but lead sulfide is practically insoluble. When exposed to the air and to hair, lead acetate reacts with sulfur to form lead sulfide, which precipitates on the hair. Proteins in hair also break down with time and release sulfur compounds, which react with the lead acetate and enhance the effect. Repeated use of such anti-graying products builds up the lead sulfide, gradually returning hair to a youthful color. At least that's what the ads say. You can usually identify people who have been using the stuff because their hair will have a dark, dull tinge. Still, many think this is better than going gray.

There is, however, one lingering concern about such products. Lead is a highly toxic element capable of poisoning the enzymes that make hemoglobin. As a result, a hemoglobin precursor called aminolevulinic acid accumulates in the body and causes toxic symptoms ranging from stomach problems to brain abnormalities. The amount of lead in these dyes is very small—less than 1 percent—and studies indicate that our blood absorbs virtually none of it. But does it contaminate the hands of those who apply it? And what about the excess lead acetate that winds up in our water supply?

Yet another problem arises when people use a permanent dye after having colored their hair with a metallic dye. Many metals, including lead, catalyze the decomposition of hydrogen peroxide into water and oxygen, and this reaction produces a lot of heat. It can actually cause scalp burns.

So there you have a summary of the hair dye saga: fascinating chemistry, and some interesting toxicological issues. But there is no question that such products make many people feel better and increase their enjoyment of life. More and more, the answer to the question of "does she or doesn't she" is a "yes." "He" is even getting in on the action, too! And since modern products can be used at home, not even the hairdresser knows for sure. But the "does she or doesn't she" question has taken on a new connotation. In the current environment of worrying about every chemical to which we are exposed, the question in people's minds is "does she or doesn't she increase her risk of cancer?" And that not even her toxicologist knows with absolute certainty.

It May Be an Alcohol—But Don't Drink the Methanol!

We take many things in life for granted. Windshield-washer fluid, for example. Only when we can't get any do we realize how important it is. Driving around peering through a dirty windshield is no fun. And it's dangerous. Thank goodness for methanol!

Water is a pretty good cleaning agent for windows. That is, as long as it stays in its liquid form. Unfortunately, since winter temperatures tend to be below freezing, we have to do something to lower the freezing point of water if we want to spray it on windshields. This is actually not that hard to do. Dissolving anything in water lowers its freezing point. Surprisingly, what is dissolved is not that important, but rather how much of it is dissolved. The number of molecules or ions present in solution is what determines the freezing point. These get in between water molecules and prevent them from coalescing to form crystals. Salt or sugar could be used, but either would be impractical. The solution would not freeze, but a deposit of the solute would be left on the windshield as the water evaporated. What is needed, therefore, is a liquid that easily mixes with water and readily evaporates. Methanol is ideal. It even has the added bonus of being a good cleaning agent. Grease, road tar, and bird droppings yield to its solvent power.

Experiments show that if we want to be able to see through the windshield down to -40°C (-40°F), we need a mix of roughly 40 percent methanol and 60 percent water. A little detergent is added, along with a blue dye. The dye serves two purposes. Market studies have shown that people generally associate this color with cleaning action (just think of toilet bowl cleaners), but more importantly, it identifies the solution as unfit for consumption. A good thing, because methanol can be deadly!

The North American public first became aware of the dangers of methanol during the Prohibition era. The Volstead Act, which operated in the US between 1917 and 1933, outlawed alcoholic beverages. Prohibitionists believed that alcohol was a major cause of crime and delinquency. The extremism of the movement was astounding in some cases. The Women's Christian Temperance Union urged schoolteachers to put a calf's brain in a jar of alcohol and show students how it turned from pink to gray. The kids were to be told that alcohol would do the same to their brains. Government agents regularly shot bootleggers and destroyed innocent people's property in their search for illegal booze. Not only was crime not curtailed by Prohibition, it flourished. Al Capone fitted his competitors with cement shoes while he made a fortune bootlegging. Illegal stills were everywhere, and the distribution of their produce was laced with crime. But even more disturbing was the fact that the illegal liquor was often laced with methanol. At the time, though, it was called "methyl alcohol."

The alcohol we normally consume in beverages is "ethyl alcohol." When this was made illegal, anything that had the name "alcohol" was pressed into service. Methyl alcohol had been known since the seventeenth century, when Robert Boyle discovered it as a component in the mixture of substances he obtained by heating wood in the absence of air. The term "methyl alcohol" was coined by Dumas, a French chemist, in 1834, from the Greek "methe" for wine and "hyle" for wood. Methyl alcohol was considered to be the "wine of wood." Not the kind of wine anyone would want to drink. Like ethyl alcohol, it did have an inebriating effect. But it had another effect as well. Death. A shot glass full would do it. If the imbiber was lucky, he only went blind.

Until Prohibition, though, there was not much of a motivation to adulterate beverages with methyl alcohol. There was no need. Ethyl alcohol was plentiful. But when the crunch came, methyl alcohol began to seep in to fill the ethyl void. It was readily available, since chemical plants in Europe and America were cranking it out, mostly through Boyle's original "destructive distillation" process. Methyl alcohol was used as an industrial solvent and as a raw material for the synthesis of other compounds, such as formaldehyde. In many cases, it was easier for criminals to get their hands on methyl alcohol than it was for them to set up moonshine stills. They figured that any customers who overindulged wouldn't be around to complain about the substitution anyway. During Prohibition, thousands were killed and blinded by adulterated booze. Some poor souls believed that the liquor could be made safe by filtering it through a loaf of bread. They often paid for this belief with their lives. Finally, the government stepped in. How? Officials urged chemical companies to abandon the term "methyl alcohol" in favor of "methanol." Maybe, if the name did not include "alcohol," people would be less inclined to think of it as a substitute intoxicant.

The repeal of Prohibition brought an end to the epidemic of methanol deaths, but sporadic poisonings do still occur. In Egypt, methanol, because it is cheap, is sometimes added to regular liquor to boost its alcohol content. This makes for an especially dangerous situation because the presence of ethyl alcohol postpones the effects of the methanol so that the drinker may keep drinking longer. A teacher at the American University of Cairo died after consuming a fair bit of adulterated Egyptian vodka. But we don't have to travel to Egypt to hear of such tragedies. They happen right here. Two men in Kingston, Ontario, overdosed on windshield-washer fluid. They had gone to a party where they drank what they thought was vodka punch. It was made with Kool-Aid, root beer, and the contents of a jug that a youngster had snitched from his father, thinking it was vodka. Apparently, the man had a clever little smuggling scheme going. He added blue dye to vodka and brought it across the border in windshield-washer fluid jugs, which he then stored in his garage. Tragically, he also kept real windshield-washer fluid there. That's what his son picked up by mistake. He had wanted to add life to the party, but instead served up death.

A similar event occurred in Hare Bay, Newfoundland, where a teenager died and others were hospitalized after drinking what they apparently thought was moonshine. It was methanol-based antifreeze. In order to avert such tragedies, there is talk of manufacturers adding a chemical with a distinct smell to windshield-washer fluid for easy recognition.

Today, most methanol is made by the "synthesis gas" process. Natural gas, which is mostly methane, is heated with steam to produce a mix of carbon monoxide, carbon dioxide, and hydrogen, known as "syngas." This syngas can be converted to methanol in the presence of a special zinc oxide/chromium oxide catalyst. Why is it manufactured? Because methanol has numerous uses. It can be added to gasoline to increase burning efficiency; in fact, Indy cars run on pure methanol. It can also be converted into methyl-t-butyl ether (MTBE), which is added to gasoline in some areas to provide oxygen for better combustion in order to produce less pollution. MTBE is somewhat controversial because it can get into groundwater through the leakage of underground gasoline tanks. Some people have also complained of headaches, dizziness, and nausea when it is released into the air from gasoline. It likely will be phased out, much to the concern of the methanol industry.

Shortages in windshield-washer fluid are typically due to unusually heavy demand caused by bad weather. While methanol is readily made from natural gas, windshield antifreeze producers can only bottle so much of the product. Is there a substitute? There sure is. Ethyl alcohol will do. Vodka has just about the right concentration to offer protection on the coldest days. So, you see, people who maintain that vodka is good against the cold are right.

CITIUS, ALTIUS, FORTIUS

Oh, I remember it well: the Olympics, Squaw Valley, 1960. The final game of the hockey tournament featured the US against Czechoslovakia. Incredibly, the Americans had knocked off the favored Canadians, and the Russians, and now only the Czechs stood between them and a gold medal. But going into the third period, the group of unheralded college players trailed the skilled Czechs by a score of 4-3. That's when Nikolai Sologubov, the Russians' superb defenseman, waltzed into the American dressing room and suggested that the players fortify themselves by inhaling some extra oxygen from tanks. His motive? If the Americans won, the Russians would end up with the bronze medal; if they lost, the Russians would go home empty-handed. Amazingly, the Americans scored six times in the third period for their first "miracle on ice!" Was the extra oxygen responsible?

As I recall, the next day's newspapers were filled with stories about the ingenuity of the oxygen boost. Nobody suggested that this was in any way unfair. Performance enhancement by means other than training was not yet a big issue, even though "doping" had tainted the Olympics since 1936. Just a year earlier, German scientists had isolated the male sex hormone testosterone and had shown that it increased muscle mass and aggression. There is little doubt that German athletes used it in the 1936 Berlin Olympics along with amphetamines, stimulants that had been found to ward off fatigue. By 1955, various analogues of testosterone, collectively referred to as "anabolic steroids," had been synthesized and made their way into the bodies of athletes clamoring for glory. It is hard to know how extensive such doping was back in those days, because urine tests for steroids were not introduced until 1973. Only in 1975 did the world's governing sport bodies officially ban the use of anabolic steroids. That certainly didn't mean these drugs were not being used. Detection techniques were relatively primitive, and as long as athletes didn't use steroids just prior to competition, they got away with it.

I remember marveling at the physique of East German swimmer Kornelia Ender at the Montreal Olympics of 1976. She took home an unprecedented four gold medals. She was built more like a man and even had an unusually deep voice. Steroids? Probably. Then, in 1988, the lid was blown off when sprinter Ben Johnson was caught cheating with stanozolol, an anabolic steroid, in the 100 meters, one of the Olympics' prime events. Since then we have looked warily on the Olympic motto of "Citius, Altius, Fortius," or "faster, higher, stronger," and have asked the question "with what?" It seems we have come a very long way since those American boys inhaled some extra oxygen. Now we must ask whether athletes have used growth hormone to bulk up, insulin to boost the body's supply of glycogen, a crucial muscle fuel, or if they have injected themselves with erythropoietin (EPO) to increase their production of oxygen-carrying red blood cells.

Why the need for EPO? Why not just inhale some extra oxygen? Simple. It doesn't work! The romanticized story of the American victory at Squaw Valley notwithstanding, red blood cells are already saturated with oxygen, and inhaling extra gas will be of no help. This was clearly shown in a landmark paper in the *Journal of the American Medical Association* in 1989. Researchers studied professional soccer players who breathed either room air or pure oxygen in a double-blind fashion before a period of exercise. There was no difference in performance, and the subjects were unable to identify which gas they had inhaled.

To increase the oxygen carrying capacity of the blood, the number of red blood cells needs to be increased. There are several ways to do this. Training at high altitude, where the air contains less oxygen, stimulates the body to produce more red blood cells. Living in dorms where nitrogen-rich air is pumped in to simulate the low oxygen concentration of air at high altitude also works. Then there are the shortcuts. Like "blood doping." Athletes withdraw a couple of pints of blood and reinfuse it months later, prior to a major competition, to increase their red blood cell count. Such blood doping is illegal and is detectible. Which is why athletes began to use EPO, a hormone synthesized by the kidneys that sends a signal to the bone marrow to produce red blood cells. EPO can be made via recombinant DNA technology and is widely used to treat anemia stemming from kidney disease, chemotherapy, or blood loss. It didn't take long for athletes to figure out that they could also avail themselves of this technology to boost performance. Nor did it take long for problems to crop up. Too many red blood cells increase the density of the blood, which in turn can lead to heart attacks or strokes. When the deaths of over a dozen cyclists were associated with the use of EPO in the early 1990s, the Olympic Committee banned the drug. The problem, though, was that injected EPO was difficult to detect, and reliable tests have only recently become available.

But some athletes may already be a step ahead. Pharmaceutical companies are working on a way to treat kidney patients by introducing the gene that codes for the production of EPO. Animal experiments are already under way. And I'm quite sure that there are athletes out there quite willing to become human guinea pigs. By the way, about those six American goals against the Czechs in the third period back in 1960? None of the four players who did the scoring had inhaled any extra oxygen! Natural adrenalin was the chemical at work.

IT WAS THE STRYCHNINE!

Alfred Inglethorp was up on his chemistry. He knew that strychnine sulfate was soluble in water, whereas strychnine bromide was not. When his wife's physician prescribed her a dilute solution of the sulfate as a heart stimulant, he put his knowledge to use. Alfred had plotted to marry the rich but somewhat foolish lady, induce her to make a will naming him as beneficiary, murder her, and run away with his mistress accomplice. The strychnine prescription presented him with a glorious opportunity. Mrs. Inglethorp sometimes used bromide salts to help her sleep, and it was a simple matter for Alfred to add some of these to the strychnine sulfate solution. The result was the formation of a precipitate of strychnine bromide, which settled to the bottom of the bottle. So, when Mrs. Inglethorp thought she was dosing herself with medicine, she was only pouring off the supernatant liquid. Then when she got down to the bottom of the bottle, she ingested a concentrated slurry of strychnine bromide, which proved to be lethal. How come she didn't notice the difference in consistency? I think the reason is simple. That would have ruined this wonderful Agatha Christie story!

The Mysterious Affair at Styles, written in 1920, was Dame Christie's first full-length novel. It introduced Hercule Poirot, destined to become the second most famous detective in literary history. The dapper little Belgian, who caught criminals by relying on his famous "little gray cells," rather than brawn, proved to be very adept at chemistry and figured out Inglethorp's dastardly scheme. This should come as no surprise, because Agatha Christie was trained as a pharmacist and her science was very sound! The chemistry involved in the precipitation of strychnine bromide is realistic, as is the use of strychnine as a tonic. The description of strychnine poisoning she provides is also accurate.

Just mention "strychnine," and people immediately think of "poison." Justifiably so. Of course, one person's poison can be another's drug; it all depends on the dose. Strychnine, a naturally occurring compound found in the seeds of the fruit of the nux-vomica tree, has long been used both as a poison and as a drug. Cleopatra supposedly investigated the seeds in her search for a perfect suicidal poison. She had prisoners and slaves swallow the seeds to see how quickly they would die. Death was fast enough, but Cleo was disturbed by the convulsions and distorted facial features that strychnine produced. She wanted her beauty preserved even after death and decided that the venom of an asp was the way to go. The spasms with arching of the back that Cleopatra feared so much are typical of strychnine poisoning. Indeed, it was just such a seizure that led Mrs. Inglethorp's physician to suspect strychnine as the cause of her death.

Strychnine, in tiny doses of course, has been used over the years as a heart stimulant and as a digestive aid, although with no established efficacy. Adolf Hitler suffered from chronic gas pains for which his physician, a semi-quack by the name of Theo Morell, prescribed "Dr. Koester's Antigas Pills." These contained both atropine and strychnine, although unfortunately there was not very much strychnine in the product. But there certainly was enough strychnine in the pills that Dr. Neill Cream prescribed for sickly Daniel Stott in Chicago back in the 1880s. Cream had graduated from McGill University with a medical degree in 1876, and then set up a practice in London, Ontario. His reputation took a beating when a young woman on whom he had performed an abortion was found dead near his office, presumably from an overdose of chloroform. Although an inquest did not implicate Cream, he left London and set up shop in Chicago.

In Chicago, he began to advertise a patent medicine for epilepsy, which led to a meeting with the young wife of Daniel Stott. A romantic entanglement followed. Then Stott suddenly died, some fifteen minutes after taking a pill given him by Cream. The coroner determined that the cause of death was an epileptic seizure, but the arrogant Cream, confident he had committed the perfect crime, wrote a letter to the coroner accusing the dispensing pharmacist of having added strychnine to the pills! This prompted an autopsy, which did indeed reveal the presence of the poison. Authorities, though, got suspicious when they found that Cream and Mrs. Stott had taken out life insurance on the victim. Finally the widow, perhaps feeling some remorse, testified that the two had indeed plotted to kill her husband. Vestiges of Alfred Inglethorp! Cream was tried and sentenced to life.

Under mysterious circumstances (some suggest bribery), Cream was released in 1891 for "good behavior." We next encounter him in London, England, where he becomes linked with the death of a number of prostitutes. One of these ladies, in the midst of her death throes, described how a man fitting the description of Dr. Cream had given her some "white medicine." Once again, an autopsy showed the presence of strychnine. The cavalier Cream now offered his services to the coroner to bring the criminal to justice. This aroused suspicion, and the doctor was put under surveillance. He was finally arrested when a policeman saw him leave a house where two other prostitutes were then found poisoned. A chemist then identified Cream as the man to whom he had sold strychnine. It took a jury only ten minutes to sentence the villain to the gallows.

When the trapdoor opened, Cream stunned the crowd as he shouted, "I am Jack." He never got to finish his sentence, but there has been speculation ever since that the man who was hung that day was the notorious Jack the Ripper! We'll never know. But one description of the Ripper has him sporting a horseshoe-shaped tiepin. And an old McGill class photo of Dr. Cream shows him wearing just such a pin.

Mercury—Pretty but Nasty

Let me tell you about the lady who consumed an entire paperback novel in one day. I don't mean consumed as in "read voraciously," I mean consumed as in "ate." This was unusual, because books were not her regular fare. She mostly snacked on Kleenex boxes and cigarette packages. These strange dining habits came to light when she sought medical help for chronic headaches, dizzy spells, and tunnel vision. Unable to diagnose the problem, her physician ordered a battery of tests, which revealed a high level of mercury in the blood. Where could it have come from? The physician's probing questions eventually identified the bizarre diet as the source of the mercury. At the time, in the 1980s, mercury compounds were commonly used as fungicides in the pulp and paper industry. Indeed, analysis of a Kleenex box revealed 83 parts per billion of mercury, and in a paperback similar to the one that served as a meal, a whopping 431 ppb. The lady was suffering from the effects of mercury poisoning!

"Pica," as the eating of unusual substances is known, is often a symptom of iron deficiency. This patient turned out to have a very low level of iron, which was quickly remedied with a week of ferrous sulfate therapy. Her penchant for eating paper disappeared, as did the headaches and other mercury-associated symptoms. Had the mercury toxicity not been recognized, the outcome could have been tragic. Like it was in the case of three lighthouse keepers on Australia's Rottnest Island.

This tiny island is home to Australia's first-ever rotating beam lighthouse. As in most lighthouses constructed in the 1800s, the apparatus containing the lenses that revolve around the light source is extremely heavy, and sits on a bed of mercury. This reduces friction to such an extent that the machinery can be set in motion by a mere touch. Unfortunately, mercury poisoning can also be set in motion. Mercury evaporates and can be readily inhaled. In sufficient amounts it can lead to delusions, irritability, insomnia, and depression. It can drive people to suicide. That appears to have been the fate of the first three lighthouse keepers on Rottnest Island. All three killed themselves. Was it loneliness, or was it the mercury? We'll never know, but there is enough concern about mercury contamination in lighthouses to have made the Canadian Coast Guard take action. While only 10 of Canada's 262 lighthouses still use the mercury system, there is enough residual mercury in many to make them out-of-bounds to the public.

Of course, you don't have to be a lighthouse keeper to suffer the ill effects of mercury. Having a curious ten-year-old
boy in the family can do it. In one specific instance, a member
of this species somehow acquired a small vial of mercury and
proceeded, for reasons known only to ten-year-old boys, to
splatter it all over the living room. His mother duly swept up
the mercury, discarded it, and thought the affair to be over. But
it was only the beginning. Soon the boy's fourteen-year-old
sister developed a low-grade fever and began to exhibit unchar-
acteristic aggressive behavior. She complained of painful wrists,
hands, and knees. Not only were the hands painful, but they
also showed a red discoloration and peeling of the skin. It didn't
take long for the father to begin to show the same kinds of
symptoms. The mother, on the other hand, was spared these
particular problems, but developed kidney disease. Amazingly,
the boy whose curiosity had unleashed the nightmare was
totally unaffected. The complex of symptoms suggested mercury

poisoning, which was confirmed by blood tests. Urine levels, paradoxically, showed wide swings in mercury content, casting some doubt on the validity of urine analysis for mercury toxicity. All three family members recovered, but before mercury levels in the air normalized, a new floor had to be laid, walls had to be replastered, and all furniture and carpets had to be thrown out. The boy was kept.

Mercury exposure can occur even without fledgling scientists. There is enough mercury in a thermometer to potentially cause a problem if it is broken. One such case involved a three-year-old boy who had to be admitted to hospital because of weight loss and difficulty walking. Blood tests revealed a high level of mercury, but its origin was a mystery. A toxicologist was finally sent to the family home and discovered mercury residues in the carpet in the boy's room as well as in the vacuum cleaner. Apparently, a thermometer had been broken, and the remains were vacuumed up. That was the wrong thing to do. Vacuuming mercury just spreads the vapor through the air. Spills should be picked up with an eyedropper and placed in a sealed container, which should then be disposed of as hazardous waste. Although alcohol or digital thermometers may be more costly than the mercury variety, they are certainly preferable in terms of potential toxicity. No one would ever contemplate committing suicide with these. But that is exactly what a nineteen-year-old Manhattan resident planned to do with a mercury thermometer. Actually, she planned to use eight mercury thermometers.

She broke the thermometers, got hold of a syringe, and injected herself in the upper arm with the mercury. She lay down and waited to die. But it was in vain. Apparently she failed to hit any veins or arteries and developed only a large infected blotch on her arm, which eventually led her to a hospital. A plastic surgeon removed most of the mercury, but six months later, she still showed about 150 times the normal amount of mercury in her urine. Remarkably, she had no symptoms of mercury poisoning. If she wanted to kill herself, she should have inhaled the mercury. That works. Here's the tragic proof.

Residents on a small street in Lincoln Park, Michigan, were awakened by the wail of sirens. Ambulance attendants rushed into a house and emerged with a sixty-eight-year-old man and his eighty-eight-year-old mother, both of whom were suffering from vicious nausea and diarrhea. By the time they arrived at the hospital, chest pain and labored breathing had set in. The next day the man's son and daughter-in-law were struck the same way and also ended up in hospital. This clearly seemed to be some kind of environmental problem, so investigators were dispatched to the home. They were astounded by what they found. In the basement they discovered a crude lab with a furnace designed to melt metal. It turned out that the younger man had been working for a company that manufactured dental amalgam, the mix of mercury, silver, tin, and other metals used to fill cavities. He had stolen some of the amalgam, hoping to extract and sell the silver. Mercury, he knew, was volatile, and therefore could be evaporated off by heat. Indeed it could. In doses high enough to poison everyone in the house. In spite of efforts to rid their bodies of mercury with dimercaprol, an agent that binds the metal and causes it to be excreted, all four died within three weeks. The house was demolished and the debris treated as hazardous waste.

A sailor in Louisiana was luckier. He was asked to guard a boat in dry-dock that was having its bottom replaced. It seems the man liked to dabble in chemistry and had read about some scheme to transmute mercury into gold by baking it inside a potato. Whether out of boredom or greed, he decided to give it a try with an ounce of mercury and an Idaho potato. All he got was mercury poisoning. He then had the nerve to claim damages under the Jones Act, which covers hazards to which sailors are exposed aboard boats. But the judge denied him safe harbor because "the sailor was unable to support the proposition that the practice of alchemy is within the duties of a seaman who is acting as caretaker aboard a bottomless vessel."

Feeding the Soil

A seed bursting into a plant is a marvelous process. Where does all the matter in the plant come from? Carbon, hydrogen, and oxygen, the basic building blocks of the carbohydrates, proteins, and fats that make up a plant, are available from water and from carbon dioxide in the air. But nitrogen, phosphorus, potassium, and some sixteen other elements needed for plant growth must be furnished by the soil. If the soil runs out, and is not replenished, plant growth suffers. Long before scientists provided detailed accounts of plant nutrition, ancient farmers learned through experience that crop yields decreased from year to year. Eventually they concluded that, like humans and animals, crops had to be fed.

Over 2,000 years ago, Chinese rice growers were already applying burnt animal bones to their fields, and nobody really knows when North American Indians began burying dead fish between rows of corn. We certainly do know that they taught the practice to the Plymouth settlers. We also know that George Washington fertilized more than American minds. America's first president took great interest in farming and concluded that the criteria for better crop growth were loose earth and soil "amendments." He experimented with manure, creek mud, plaster of Paris, lime, "green manure" (plowing buckwheat, clover, and peas into the soil), and fish heads! The president was certainly on the right track; each of these "amendments" was capable of contributing some nutrition to the soil.

The most significant advance in fertilizer development, however, came about in an accidental fashion. And for this we can thank a Spanish missionary to Chile whose name has been lost to history. It seems some native Indians had extinguished a fire by throwing hard, dry earth onto the hot coals, and were stunned by the acrid purple vapors that were suddenly released. Some sort of evil spirits, they probably thought, and grabbed a few chunks of the dry earth to show the priest, whom they assumed would have an explanation. He didn't, and threw the samples into his garden. A few months later, the alert missionary noted an increased growth of vegetation where the chunks had landed. "Chile saltpeter" had made a triumphant entry into the world of agriculture. As chemists later learned, it was mostly sodium nitrate, an excellent substance for introducing soluble nitrogen into the soil. Nitrogen is a component of all proteins, including the enzymes that are instrumental in every phase of plant growth. Many of the vitamins that plants produce, as well as the chemicals with which they protect themselves against insects and fungi, contain nitrogen. But what caused the irritating, purplish vapors? Saltpeter is commonly contaminated with sodium iodate, which, when heated, releases a variety of iodine compounds that can cause a choking sensation.

By the seventeenth century, Chilean saltpeter was widely imported into Europe, along with another South American commodity that had been found to increase soil fertility. This was guano, or bird excreta. Like saltpeter, it was an excellent source of nitrogen and furnished phosphorus to boot. Peru was the first recognized source of extensive deposits of bird poop, but by the 1800s, many South Pacific islands were also found to be overflowing with seabird guano. To this day, guano is an important source of fertilizer, helping to feed the world. But it has also helped to overfeed the natives of the guano-producing islands. Contrary to what many may think, Americans are not the fattest people in the world. That dubious honor goes to the inhabitants of the Pacific island of Nauru, who boast one of the highest per-capita incomes in the world. And that thanks to bird droppings! Fertilizer companies pay high prices for the chance to harvest the guano, allowing natives to trade in their plowshares for easy chairs. As their level of activity decreased, imports of beer, meat, and chips increased. The result is that about 70 percent of the natives of Nauru are obese, and a third suffer from diabetes.

Exactly why saltpeter and guano increased yields interested the great German chemist Justus von Liebig. He decided to solve the mystery by burning plant material and analyzing the residue. In 1840, he published his classic, *Organic Chemistry in Its Applications to Agriculture and Physiology,* which clearly established him as the father of modern soil science. Liebig's analysis revealed that the major minerals present in plant residue were nitrogen, phosphorus, and potassium, and that the reason saltpeter and guano enhanced plant growth was because they were rich sources of nitrogen and phosphorus. Potassium, he said, could be supplied by potash. This was originally obtained by soaking the ashes of wood to dissolve the potassium salts, then filtering the suspension and boiling off the water in a pot to leave a white ash, "potash," which was mostly potassium carbonate.

Liebig's theories launched research into the systematic development of mineral fertilizers, without which we could not possibly feed the population of the world. And without these fertilizers, we would not have the abundance of fruits and vegetables with which we are blessed today. Still, some look on "chemical" fertilizers with a wary eye and favor crops grown with the aid of "organic" fertilizers, such as manure. Now, I have nothing against animal or human dung, but the idea that it can feed the world is, well, romanticized bull manure.

Technology generally arises in response to a need. And that was the case with chemical fertilizers. They were born out of necessity, simply because the traditional methods of fertilizing the soil via crop rotation or the application of compost or manure were not getting the job done. By the middle of the nineteenth century, agricultural yields in Europe had declined so dramatically that famine was in the offing. Large-scale tragedy was averted with the discovery that food production was limited by the soil's nitrogen content. Unfortunately, fertilization techniques being used at the time were just not replenishing the nitrogen absorbed by crops. Enter sodium nitrate, or "Chile saltpeter," an excellent nitrogen source. It's hard to see why the application of this "chemical" substance to fields should be regarded as any less natural than the application of manure; after all, saltpeter is the end result of the decomposition of animal waste, and is mined from the earth. What could be more natural?

Chilean saltpeter deposits were not enough to satisfy the needs of a rapidly growing population, and the problem of supplying the soil with enough nitrogen wasn't solved until Fritz Haber found a way to make ammonia from nitrogen and hydrogen. Ammonia could then be converted to ammonium nitrate, an ideal nitrogen fertilizer. What an irony that plants are surrounded by a vast amount of nitrogen (the gas makes up 80 percent of air), but are unable to use it!

Plants can only absorb soluble nitrogen compounds from the soil. Some, the legumes, harbor specialized bacteria in nodules on their roots that are able to "fix" nitrogen from the air. In other words, they can convert nitrogen gas into nitrates, which the plant can use directly for growth. That's why legumes can be plowed into the earth as "green manure," and why they serve a vital role in crop rotation. All other plants have to rely on soil microbes to convert nitrogen containing organic compounds into soluble nitrates. Such compounds, urea being an example, can be found in manure. Or nitrates can be supplied directly by fertilizer. Basically, as far as a plant is concerned, whether the nitrates it needs are supplied by manure or by the addition of a chemical fertilizer is irrelevant. It is not irrelevant, however, as far as the soil is concerned. Manure is better for soil structure; it contains substances such as humic acids, which limit nitrate leaching, help retain water, and reduce erosion. Chemical fertilizers are more likely to allow nitrate to be leached out into water systems where they can fertilize aquatic plants, which eventually die, decompose, and use up some of the dissolved oxygen needed by fish for their survival.

There are other concerns with the application of mineral nitrates. Overuse leads to their buildup in the soil, where bacteria can convert nitrates to nitrous oxide—a "greenhouse gas" hundreds of times more potent than carbon dioxide. Fertilizer producers have addressed these issues and have developed a number of "slow release" fertilizers. For example, synthetic urea can be converted to urea-formaldehyde, which releases nitrogen gradually through microbial activity in the soil. There are also granular fertilizers, coated with semi-permeable membranes for slow release, or encapsulated in microcrystalline wax. Fertilizers today can be formulated with virtually any ratio of nitrogen, potassium, and phosphorus, the main nutrients plants require, and can be matched to a particular soil's needs. Manure has a more random composition and is much lower in nutrients, particularly phosphate. Applying it to the land is more difficult and more expensive. It's not that organic agriculture based on manure cannot work; it can. There is no doubt that in test plots, or on specific farms, organic farming can be very effective. Indeed, experiments in test plots have shown that using manure as fertilizer can match the yields produced by mineral fertilizers, and that the soil is less prone to nitrate leaching.

And without a doubt, there are even some large and successful organic farms.

The Pavich family in California farms about 1,800 acres and produces 12,000 tons of table grapes a year using only composted steer manure as fertilizer. But this success could not be duplicated by rice growers in India, or by wheat farmers in Africa. There isn't enough manure available locally, and transportation costs would be prohibitive. Even when sufficient manure is available, problems still crop up. Numerous cases of food poisoning caused by *Salmonella* or *E. coli* bacteria have been related to the use of manure as fertilizer. *E. coli* O157:H7, the bacterium in contaminated tap water that sickened over 2,000 people and killed seven in Walkerton, Ontario, in 2000, can survive in bovine feces for seventy days. In fact, it probably entered the water system from manure. But even if manure is properly composted to eliminate bacteria, the bottom line is that it is not going to solve the global hunger problem. Without the use of nitrogen fertilizers, we could perhaps feed about half the world's population. Indeed, it is hard to think of a scientific development that has had a greater beneficial impact on human life than the often-maligned "chemical fertilizers."

AMMONIUM NITRATE—MORE THAN A FERTILIZER

It was early in the morning on April 16, 1947, but spectators flooded to the docks in Texas City, Texas, nevertheless. They were drawn to see the bright orange flames and the massive plume of black smoke that enveloped the ss *Grandcamp*, a French ship that had caught fire in the harbor. Then, as people marveled at the inferno, and quick-thinking vendors circulated with peanuts and other refreshments, there was a reverberating explosion. Hot pieces of metal from the disintegrated ship rained down, a devastating shock wave rolled across the land and sea, and within minutes, much of Texas City was in flames. Almost 600 people perished, many of them the onlookers who had come to gawk at the spectacle.

What cargo was responsible for the disaster? Nitroglycerine? TNT? Dynamite? No, it was none of the above. It was fertilizer! Not any old fertilizer, mind you. The *Grandcamp* had been loaded with 2 million kilos of ammonium nitrate destined for Europe. Ammonium nitrate is rich in nitrogen and can yield bumper crops or green up a lawn. But it can also figure in an explosion and cause terrible bloodshed. An explosion can best be described as a "sudden going away of things from the place where they have been." The cause of such swift departures is a shock wave formed by the very rapidly expanding gases that characterize an explosion. In the case of ammonium nitrate, the gases are water vapor, oxygen, and nitrogen. Don't get the impression that ammonium nitrate explodes easily, though. It doesn't. The chance that the bag of ammonium nitrate fertilizer you may have purchased from a garden supply store will spontaneously blow up is roughly zero. Various conditions have to be met for an ammonium nitrate explosion to occur.

Let's get back to the ss *Grandcamp*. A fire broke out in the hold, most likely due to an improperly discarded cigarette. Fearing damage to his cargo, the captain decided not to try to extinguish the flames with water. Instead, he ordered the hatches to be battened down, hoping to cut off the fire's oxygen supply. It didn't work, and the cargo of ammonium nitrate began to heat up. At first, it just decomposed into steam and nitrous oxide, better known as "laughing gas." But it was no laughing matter when the high temperatures triggered the breakdown of the laughing gas into nitrogen and oxygen. The fire, now well supplied with oxygen, intensified. Still, there probably would have been no explosion had it not been for two other factors: the ammonium nitrate was packed in paper bags, which began to burn with great intensity, and, more significantly, the ship had been filled with 1,500 tons of fuel oil. When the oil caught fire and its hot vapors mixed with the ammonium nitrate, which by now was venting massive amounts of oxygen, the conditions were right for "a very sudden going away of things from the place where they had been!"

The Texas City disaster was an accident. But the 1995 bombing of the Federal Building in Oklahoma City certainly was not. The chemistry, however, was the same. The perpetrators of this horrific crime were aware of the explosive nature of mixtures of ammonium nitrate and fuel oil (ANFO). Indeed, most commercial explosives used for mining and construction in North America fall into this category. ANFO mixtures are actually remarkably safe to use; they have to be detonated by an explosive charge. Of course, Timothy McVeigh and his cronies had no access to commercial explosives, so they decided to make their own. Either they researched their subject remarkably well, or they were very lucky. Homemade brews of fertilizer-grade ammonium nitrate and fuel oil are very difficult to detonate without the use of TNT, dynamite, or blasting caps. Unfortunately, as history has shown, terrorists can be remarkably resourceful. That's why chemical companies and the fertilizer industry are examining ways to prevent the use of ammonium nitrate fertilizer from being used as an explosive.

The first efforts along these lines were taken in Northern Ireland and the Republic of Ireland, where the government decreed that all ammonium nitrate sold as fertilizer had to be mixed with limestone (calcium carbonate). This reduces the explosive potential, but does not eliminate it. Many European countries also have laws stipulating that ammonium nitrate destined for fertilizer use has to be manufactured in such a way as to limit oil retention and must also be very low in carbon, chlorine, and copper contaminants, as all of these increase the sensitivity for explosion. In the US, as early as 1968, patents were issued for a combination of ammonium nitrate with diammonium phosphate, which supposedly greatly reduced the chance of detonation. In fact, after the Oklahoma City disaster, four victims filed lawsuits against a fertilizer producer, claiming that the bombing could have been prevented if the appropriate additives had been used. The suit was eventually dismissed, but the company in question did carry out tests to investigate the potential of such additives. Unfortunately, the results showed that while they worked on a small scale, they were ineffective on a scale that terrorists would use. Currently, the thrust is to develop a polymer coating for fertilizer granules that prevents oil absorption. This coating is designed to dissolve in soil, so it should not impair fertilizer activity. Now perhaps you'll understand why I and other chemists get nervous when we hear of large amounts of stored fertilizer disappearing, as has recently happened in England, Australia, and Thailand. I hope green thumbs, not bloody hands, are at work here.

CHLORATE AND THE EXPLODING TROUSERS

My high school chemistry classes were not impressive. We memorized tons of formulas and drew loads of pictures of experimental setups. We became very adept at drawing condensers, Erlenmeyer flasks, and Bunsen burners with a template, but rarely did we get a chance to actually perform an experiment. But there were a few. I do recall heating a mix of manganese dioxide and potassium chlorate to produce oxygen, which we collected and tested with a glowing splint. The ability of oxygen to support combustion was driven home as the splint burst into flame. Potassium chlorate, $KClO_3$, we learned, could liberate its oxygen content when heated in the presence of manganese dioxide, which served as a catalyst. It was an "oxidizing agent." I was so impressed that I snitched a bit of the chlorate to experiment with at home, carrying it in my pocket. Not a smart thing to do, as you'll discover.

Although I don't think I realized it at the time, our experiment was very similar to the one Joseph Priestley performed in 1774, which led to the discovery of oxygen. Priestley heated mercuric oxide, HgO, and collected the gas produced. "What surprised me more than I can well express, was that a candle burned in this air with a remarkably vigorous flame...." Priestley, though, didn't recognize the gas as an element, calling it "dephlogisticated air," in light of the prevailing belief that flammable materials contained phlogiston, a substance without color, odor, taste, or weight, given off during combustion. "Phlogisticated" substances were those that contained phlogiston and, when burned, produced "dephlogisticated" air, which is what Priestley thought he had isolated.

Upon inhaling this air, he wrote: "The feeling of it to my lungs was not sensibly different from that of common air; but I fancied that my breast felt particularly light and easy for some time afterwards. Who can tell but that, in time, this pure air may become a fashionable item of luxury." Priestley was right; salons where people go to breathe oxygen have cropped up, usually making nonsensical claims about the benefit of this practice. Amazingly, he had some pertinent observations about the possible hazards of breathing "dephlogisticated air." "We may infer from these experiments that though pure dephlogisticated air might be very useful as a medicine, it might not be so proper for us in the usual healthy state of the body, for, as a candle burns out much faster in dephlogisticated than common air, so we might, as may be said, live out too fast, and the animal powers be too soon exhausted in this pure kind of air."

Today we realize that oxygen of course is necessary for life, but inhaling it also results in the production of free radicals, those rogue molecular species that can wreak havoc in the body. Antioxidants in the form of certain vitamins and plant components can curb the destructive effects of oxygen, which account for the attention they receive in both the popular and scientific literature.

Although Priestley usually gets credit for the discovery of oxygen, he certainly was not the first person to produce the gas. Michael Sendivogius, a Polish alchemist, found back in 1604 that heating saltpeter (KNO_3) produced what he called "the elixir of life." Some historians even believe that Cornelis Drebbel, who, using wood and greased leather, designed the world's first submarine in 1621, explored the possibility of heating potassium nitrate to supply his crew of twelve oarsmen with breathable air. In Sweden, two years prior to Priestley's experiment, Carl Wilhelm Scheele produced oxygen from mercuric oxide, recorded his observations, but did not publish them until several years later. Priestley, on the other hand, carefully documented his work and published his results promptly. Antoine Lavoisier correctly interpreted Priestley's experiment as having produced a new element, but did not attribute much credit to the Englishman. Lavoisier maintained this attitude despite historical evidence that his own experiments with oxygen had been prompted by a meeting he had with Priestley. When taken to task on this issue, Lavoisier commented that "those that start the hare do not always catch it."

While there is controversy about who discovered oxygen, there is no doubt that it came about by heating some oxidizing agent. This fueled the idea that oxidizing agents could enhance combustion. Potassium chlorate, for example, made matches and sparklers possible. Other uses were discovered as well. Sodium chlorate proved to be a good source of oxygen for combining with chlorine to yield chlorine dioxide, a better bleaching agent for paper than plain chlorine. Amazingly, sodium chlorate even turned out to be an effective weed killer. But this weed killer had an occasional bizarre side effect. It caused farmers' pants to explode!

Ragwort was a hugely problematic weed in fields in New Zealand where dairy cows grazed in the early years of the twentieth century. Some of the alkaloids the plant contains can cause liver failure and kill cattle. So it was with relish that farmers began to spread sodium chlorate on their fields after learning that the chemical was effective in destroying ragwort. What they didn't realize was that sodium chlorate was a strong oxidizing agent that, when combined with combustible materials like cotton or wool, could cause violent explosions. Even after laundering, pants that had been exposed to chlorate could still hold enough of its residue to produce dramatic effects. In one widely reported case, a farmer's pants were drying in front of a fire, and exploded with a loud report. He had enough presence of mind to hurl the remnants out of the house, where "they smoldered on the lawn with a series of minor detonations." I guess this could have happened to the pants I had on when I snitched the chlorate. Luckily, we didn't dry our clothes by hanging them in front of a fireplace.

THE MUSIC OF COPPER SULFATE

You can imagine that an album entitled *Copper Sulfate Crystals,* recorded by "Man in Formaldehyde," would capture my attention. What could this be all about? Was some eccentric musical genius inspired by an experiment gone awry in a chemistry lab? I had to find out, especially since I've always had a fondness for those beautiful blue crystals. I've admired them many times in the lab, but I've never had the chance to listen to them.

Actually, another experiment I remember from high school (besides heating potassium chlorate—see above) involved copper sulfate. I remember attaching the leads from a battery to two graphite pencils immersed in a copper sulfate solution, and watching in amazement as the pencil tips became coated with metallic copper. The copper ions in the solution had picked up electrons from the battery and were deposited as copper atoms on the graphite. But I remember something else as well. I remember Mr. Cook warning us to take care with copper sulfate, because it could be toxic if misused. That was somewhat of a revelation because I remembered that as a youngster, I used to grow pretty crystals by hanging a thread into a solution of copper sulfate. I didn't recall that the chemistry set I used came with any such warnings. Of course, that may have been because I never read the instructions. Nor did I realize at the time that these crystals had a fascinating history going back all the way to the ancient alchemists.

Copper sulfate occurs in nature, with "chalcanthite" being a particularly attractive form of the mineral. I'm not surprised that the alchemists found its blue luster alluring. Unlike most minerals, chalcanthite is quite soluble in water, a property that lends itself to experimentation. Somewhere along the line, an alchemist discovered that immersing a piece of iron in a solution of copper sulfate resulted in a dramatic effect. The iron seemed to have turned to gold! Of course, that is not what had really happened. The copper ions had stolen electrons from the iron and had deposited on its surface as metallic copper. One wonders how many "clients" were taken in by this "transmutation."

Copper sulfate–induced folly was not limited to the alchemists. In 1891, Dr. Varlot, a French surgeon, developed a way to copper-plate a corpse for preservation. The body was dipped into silver nitrate, then placed into an evacuated chamber, where it was exposed to vapors of phosphorus that reduced the silver ions to metallic silver. Then came immersion in a copper sulfate solution. Since silver, like iron, can donate electrons to copper ions, Varlot was soon gazing at a body electroplated with copper. Why did he engage in this bizarre practice? It seems the good doctor had some ideas about preserving bodies for later resuscitation. This was pretty implausible, especially after the corpses had been exposed to such large amounts of copper sulfate. Like Mr. Cook told us, the stuff really can be toxic. That's why it is used as a fungicide on grapes, as an algae-cide in swimming pools, and is thrown into pig manure pits to deal with the bacteria that produce those noxious smells.

All of this doesn't mean that we should not allow students to carry out experiments with copper sulfate. But those experiments have to be prefaced with the appropriate warnings. Unfortunately, though, such warnings are sometimes taken the wrong way. And that is just what seems to have happened in a high school in Sylvan Lake, Alberta. Apparently, three girls had taken a dislike to a classmate and decided to have some fun at her expense by stealing some copper sulfate from the school lab and mixing it into a "slushie" they had purchased at a convenience store. I'm not sure why they chose copper sulfate, but a good guess would be that they recalled some sort of warning from the teacher when they were using it in an experiment. The victim consumed the beverage and got sick. So did two of the pranksters, who sipped a little of the slushie to convince the suspicious victim that there was nothing wrong with it. Four other girls also somehow drank from the spiked beverage and experienced a variety of symptoms, including vomiting, shaking, headaches, and a burning sensation in the mouth. They were treated at a hospital and released.

Copper sulfate can be deadly, but chances are that any significant ingestion would trigger vomiting, and thereby expel most of the dose. That is just what happened in the Alberta school. The three culprits didn't know that this would happen, and were charged with attempted murder. This was recently plea-bargained down to "administering a noxious substance with the intent to endanger life, theft of copper sulfate, and criminal negligence." The potential penalty here is less severe, but the girls could still be looking at jail time.

Now back to my "Man in Formaldehyde" recording. I had to hear what this *Copper Sulfate Crystals* music was all about. I downloaded a little excerpt from the Web (legally), and with some trepidation, I began to listen. After all, some of this modern stuff that passes for music is pretty lethal. What a pleasant surprise! It seemed to be some sort of mix of guitar and computer-generated tunes that were melodious and pleasant. What it has to do with copper sulfate, I have no idea. And I don't know what the other tracks sound like either. But I will. I sent for the CD, of course. How could I not want to listen to "Birds in Magnetic Milk"?

FLYING HIGH WITH ALUMINUM

The UFO hovered high in the sky as the TWA pilot approached the Long Beach Municipal Airport in California. As he got closer, he couldn't believe his eyes. There, at 16,000 feet, sat a man in a lawn chair held aloft by dozens of balloons! He held a gun in his lap and had a parachute strapped to his back. Was this some kind of novel terrorist activity? No, it was Larry Walters, an adventurer who had planned to balloon across the Mojave Desert in an aluminum lawn chair attached to forty-five weather balloons filled with helium. The gun? That was his landing gear. Larry had planned to come back to earth by bursting his balloons with pellets! Unfortunately, the landing was not quite as soft as he had hoped. He got entangled in power lines, and firefighters had to come to his rescue. His punishment was a fine for operating a civil aircraft with no "air-worthiness certificate."

Actually, Larry had planned the air-worthiness of his craft quite carefully. That's why he chose an aluminum lawn chair. He knew that the metal was durable and extremely light! In fact, the pilot who spotted the strange flying contraption was himself at the controls of an aircraft made largely of the same metal. Aluminum is ideal for such uses. Besides being light, it doesn't corrode easily, and it can be economically produced. Jules Verne was one of the first to recognize the potential of aluminum in flight. In his classic work *From the Earth to the Moon,* written in 1865, he described aluminum as "easily wrought, very widely distributed, forming the basis of most of the rocks, three times lighter than iron, and seems to have been created for the express purpose of furnishing us with the material for our projectile." Verne certainly was a visionary. The first satellite to be launched into earth orbit, the Soviets' *Sputnik,* was made of aluminum, as is much of the Space Shuttle. Back in 1865, though, the construction of such a large object out of aluminum was just a figment of the imagination.

Aluminum was certainly known at the time, and it wasn't rare. Actually, it is the most abundant metal in the earth's crust. But the metal is not found in its elemental state; it only occurs in combination with other elements. Clay consists of aluminum silicates, and most of the rocks in the world contain aluminum. Bauxite, named after Les Baux in France, where huge deposits were first found, is basically made of aluminum oxide. Separating aluminum from the other elements defied scientists until 1827, when Friedrich Wohler, in Germany, managed to tease a few bits of aluminum out of aluminum chloride by reacting it with potassium in a platinum crucible. Still, aluminum remained a laboratory curiosity until the middle 1800s, when Henri Sainte-Claire Deville found a way to isolate the metal by passing an electric current through aluminum chloride fused with sodium. This was not a commercially viable process, but it did make aluminum available in small amounts. As long as someone was willing to pay the price! Aluminum was judged to be more precious than gold at the time, and at the Paris Exposition of 1855, a small ingot was exhibited next to the crown jewels. Cost, it seems, was no impediment to Emperor Louis Napoleon III, who ordered a set of cutlery made entirely of aluminum. But it was not for everyday use. The aluminum utensils were rolled out only for state occasions. On a daily basis, members of the royal household had to make do with gold cutlery!

Americans were not to be outdone by the trappings of the French court. When the Washington Monument was constructed in 1884, architects searched for a unique way to top it off. It was decided that a pyramid made of aluminum would be a fitting crown for the monument that was destined to become, and remain, the tallest structure in the city of Washington. The little pyramid weighed only about 6 pounds, but at the time was the largest piece of aluminum that had ever been cast.

Then, in 1886, along came Charles Martin Hall. He had just graduated from Oberlin College in Ohio. His chemistry professor at Oberlin, Frank Fanning Jewett, was a former pupil of Wohler's, and had often regaled his students with stories about Wohler's attempts to produce aluminum on a large scale. A fortune would await anyone, he said, who solved this problem. At the tender age of twenty-two, Hall did exactly that! Using a homemade battery, he discovered that electrolysis of a solution of aluminum oxide dissolved in cryolite (sodium aluminum fluoride) resulted in the release of oxygen at the positive electrode as molten aluminum collected around the negative electrode. The process was relatively easily scaled up, so that by 1890, aluminum was available for 60 cents a pound. Curiously, the same year that Hall made his discovery, Paul-Louis-Toussaint Heroult, who was also twenty-two years old, came up with the same process completely independently, in France. A further bizarre quirk is that both men were born in December 1863, and died at the young age of fifty-one in December 1914.

The Hall-Heroult process quickly converted aluminum from a precious metal to a commodity. By 1893, when the aluminum Statue of Eros was unveiled in London's Piccadilly Circus, the metal was no longer a rarity. Aluminum foil was manufactured in France as early as 1903, the same year the Wright brothers made their first historic flight. And that flight would not have happened without aluminum! The airplane could accommodate only 200 pounds for the engine, and the only material light enough was aluminum. The Wrights, like Larry Walters, knew that if they wanted to fly, they needed aluminum. The brothers, of course, went down in history. And Larry? He paid his $1,500 fine and went on the talk show circuit. He even made a Timex commercial featuring an adventurer's watch. You guessed it—the watch casing was made of aluminum.

BLUE GARLIC AND GOLD SMUDGE

It isn't exactly an earth-shaking problem, but you would think someone would have solved it by now. Why does pickled garlic sometimes turn blue? People get unnerved when they pick up a jar in which some of the cloves have a blue tinge. Surely, it must be spoiled, they think. Well, no. We've known for a long time that blue garlic, while it may seem unappetizing, is completely safe to eat. The puzzle has been why cloves turn blue, and why it only happens to some cloves, in some jars.

Believe it or not, the blue garlic question has been investigated for over fifty years, but it was only recently that a paper published in the *Journal of Agricultural and Food Chemistry* finally got to the heart of the matter. Oh, it isn't that explanations haven't been offered before. They have. But they've been wrong! Two basic theories have been advanced in the scientific and the popular literature. One held that the discoloration was due to the buildup of copper sulfate. Not an unreasonable idea, given that copper sulfate is indeed blue, and that copper can be present in our water supply. The theory was that sulfur compounds known to be present in garlic could convert to sulfate, which then reacts with copper in water to form the colorful copper sulfate. Since copper concentrations in water vary, the problem wasn't expected to occur all the time. Sounds reasonable, but this explanation is wrong. Experiments can readily show that doping water with extra copper before immersing garlic cloves does not necessarily produce the blue color.

The other theory advanced was that garlic contains compounds called anthocyanins, which change color depending on the acidity of the surroundings. While anthocyanins in red cabbage may be susceptible to such changes, those in garlic are not. It is simple enough to show that pickling garlic with different amounts of vinegar—that is, different amounts of acid—does not correlate with blue color formation.

So what, then, is the answer? As a team of experts from the Czech Republic and the US found, the discoloration is due to pigments that form between sulfur compounds in garlic and amino acids. Isoalliin is found in garlic, and when garlic tissue is disrupted, as happens in processing, an enzyme is liberated and reacts with it to form thiosulfinates, compounds that then react with the garlic's naturally occurring amino acids to form blue pigments. The age of the garlic determines how much isoalliin there is in the first place, and the nature of the processing determines how much enzyme is liberated. Some of the mysteries still have to be cleared up, but you can rest assured that there is no harm in consuming your blue garlic. The pigments that form by the reaction of the thiosulfinates with amino acids are not toxic.

So that mystery is solved. But another color conundrum that I've been asked about over the years remains. Try this. Take a gold ring and slowly rub it across your cheek or the back of your hand. Some of you will see a black line, just as if you had used a pencil on the skin. This is known as "black dermographism," which translates literally to "black writing on the skin." In common language it is usually referred to as "gold smudge."

Now, if you are looking for a little fame and fortune, just find the cause of this problem. There are all sorts of theories, but not one of them can explain all cases. This may be because not all cases have the same cause. Let's start by listing the facts. The effect has been noted with all gold jewelry, irrespective of karat value. Women experience it more than men, and more if they are wearing makeup. Some women claim they see it more at certain times of the month, or if they are stressed, or if they have been eating acidic foods.

There is no doubt about the connection to cosmetics. Any finely powdered metal will appear to be black, and cosmetics contain abrasives that can act like fine sandpaper on gold. Many cosmetics contain zinc oxide, titanium oxide, calcium carbonate, and iron oxide as pigments. Both titanium dioxide and zinc oxide are harder than gold or silver, so as the gold is drawn across the face small amounts will abrade and discolor the skin. twenty-four-karat gold makes more intense marks because it is the softest form of gold and is the most readily abraded.

But what happens with men or women who don't wear any makeup, and still experience gold smudge? We may be looking at a different cause here. Gold jewelry is not pure gold; the metal is alloyed, usually with copper and silver. Both of these metals can undergo a chemical reaction with sulfur compounds in the skin, stemming from the breakdown of sulfur-containing proteins. Both copper sulfide and silver sulfide can form black marks on the skin. The breakdown of the proteins in question may be hormonally controlled, which might explain why some women experience gold dermography more noticeably at certain times of the month. These reactions are more likely when the skin is more acidic, since copper and silver from the ring then become more soluble.

I have searched the scientific literature, and it seems that nobody has engaged in a systematic study of the causes of gold smudge. It is high time that science paid attention to such important issues, and I would welcome offers from prospective scientists to take up this challenge. If you have any comments, or any theories about gold smudge, please let me know. I, for one, intend to start testing everyone I come across. So if you see people with black marks on their faces, you'll know that I'm on the job.

A STABLE MASS OF BUBBLES

We wash with them. We shave with them. We shape our hair with them. We sit on them. We put fires out with them. We drink from them. We even eat them. What are they? Foams! They are of both scientific, and mythological, interest. According to legend, Aphrodite, the ancient Greek goddess of love, was born from the white foam ("aphros," in Greek) produced when the severed private parts of the god Uranus were tossed

into the sea. Uranus is central to the Greek creation myth, but this god apparently had a real character flaw. He hated his children and hid them from view once they were born. One of them, Cronus, objected to this treatment and sought revenge by castrating his father with a sickle. He then triumphantly flung the severed parts into the ocean! And it was from this bloody foam, as the story goes, that Aphrodite was created. I don't know about that, but the sea certainly does foam. We'll get back to that after we find out what foams are all about.

Simply stated, a foam is nothing more than a stable mass of bubbles. The most common variety forms when a gas is dispersed in a liquid. Shaving cream and hair mousse are typical foams. They consist of tiny pockets of gas surrounded by a thin film of water. Critical to the formation of a foam is the ability of water to form a stretchable film. Pure water will not foam because water molecules are attracted to each other very strongly. This attraction is called surface tension, and must be reduced in order to form a foam. Dissolved proteins can do this by getting in between the water molecules. That's why egg whites, which contain plenty of proteins, can be foamed. Furthermore, as these proteins come into contact with air during whipping, their long molecules unfold and bond to each other, strengthening the bubbles and preventing them from collapsing. That's the basis of meringue.

You can even make a meringue without an eggbeater. Just dilute some egg white with three times as much water, and add baking soda. Then add some citric acid. Carbon dioxide will form as the bicarbonate reacts with the acid. The gas rises to the surface, but does not break through and escape into the air like it would in a pure liquid because the surface layer of water can stretch to accommodate the gas and form a bubble. As it stretches, the protein molecules coagulate and help form a tough stable film or, in other words, a bubble. As more gas rises to the surface, more bubbles form, and pretty soon we have a foam or, in French, a "mousse."

One of the great advantages of foams is that they allow for the dispersal of small amounts of chemicals over a large area, as in the case of a hair mousse. When we mousse, we actually coat the hair with a thin layer of plastic. We certainly don't want a lot of plastic in our hair, just enough to hold it in place. Remember that a mousse is mostly gas, with a little liquid stretched around each bubble. The active ingredient is dissolved in this little water. Basically, a mousse makes a small amount of liquid, and whatever is dissolved in it, go a long way. Shaving cream is the same idea. This time, the substance dissolved in water is not a plastic; it's soap. But again, a little soap goes a long way. And here's an interesting way you can make use of this notion. Since shaving cream contains little water, it is ideal for cleaning jobs where water is undesirable, such as on upholstery.

Another practical use of foams is in fire extinguishers. A fire is sustained only when there is ample fuel, oxygen, and heat, and it can be extinguished by removing any one of these components. Water removes heat as it vaporizes. But water can actually spread an oil fire, since oil floats on top of water. It cannot be used on electrical fires, either, because water conducts electricity and can cause electrocution. Even when the use of water is appropriate, it can cause extensive damage. One way to minimize the amount of water is through the use of carbon dioxide foam fire extinguishers. Such foam acts as a wet blanket on a fire, preventing the burning material from contacting oxygen. A carbon dioxide foam is made by combining solutions of sodium bicarbonate and aluminum sulfate, which are stored in separate compartments in the fire extinguisher. When mixed, these chemicals produce carbon dioxide as well as aluminum hydroxide, which is a great foaming agent.

A foam does not necessarily have to contain water; it is just a stable mass of bubbles. A foam rubber pillow, or foamed polyurethane insulation, or a Styrofoam cup all consist of bubbles surrounded by various flexible materials. To make foam rubber, for example, ammonium carbonate can be mixed with the rubber latex. When this is heated, carbon dioxide and ammonia are produced, and become entrapped in the complex rubber structure.

The type of substance added to reduce the surface tension of water in order to make a foam is critical. If it is too effective at reducing surface tension, the bubbles will burst because the attraction between water molecules is too greatly weakened. When this happens, we have an anti-foaming effect. Anti-foaming agents, typically silicones, are often added to oils to prevent bubbling. They are also used to prevent large bubble formation in our digestive tract. Simethicone is a typical anti-flatulent given to gassy people. This ensures that intestinal gases escape in a slow steady stream instead of in large bursts.

Foam is not welcome in our digestive tract, but it is welcome in beer. Indeed, foaming agents are sometimes added on purpose. Propylene glycol alginate is an example. It can be added to light beer, which has fewer carbohydrates capable of stabilizing the foam. And where does one find the alginate used in beer? In seaweed. And now you understand why the sea foams. It has absolutely nothing to do with Uranus' private parts.

The Paper Trail

It was a quasi-religious moment. There in front of me, in a display case at the British Museum, lay the original copy of *The Adventure of the Missing Three-Quarter*, in Sir Arthur Conan Doyle's own hand. Like any other Sherlock Holmes fan, I have read and reread the detective's adventures numerous times, but never before had I gazed upon an original version. Unfortunately, the hallowed moment was a little tainted by the aged appearance of the manuscript. It was a brownish yellow in color! Of course, one would expect a 100-year-old piece of paper to show its age. That was no surprise. But the appearance of the *Missing Three-Quarter*'s neighbor was. A Gutenberg Bible, produced over 500 years earlier, looked as good as new! And it will likely be on display long after the Sherlock Holmes manuscript has crumbled away along with millions of other books stored in the British Library and other major libraries around the world. What is the difference? It all boils down to the paper that was used.

Ah, paper. We don't give it much thought, but our society would grind to a halt without it. Remember those promises that computers would provide a "paperless society?" Forget it. We use more paper than ever. Rough copies spew out of our printers, and we use reams of paper to feed our Internet habit. Yet most people have no idea of the complex chemistry involved in producing the marvelous product that gives us grocery bags, facial tissues, toilet paper, books, and a myriad of other products, including newsprint.

The earliest forms of paper were not that complicated. Thousands of years ago, the Egyptians scraped out fibers from the inside of the bark of the papyrus plant (our word *paper* derives from this), and pressed them into sheets. Actually, though, papyrus wasn't really paper. Not by our modern definition, anyway: paper is the substance that forms when a slurry of disintegrated cellulose fibers is allowed to settle on a flat mold. When the water is drained away, the deposited layer can be dried into paper. The oldest surviving such piece, although devoid of any markings, was discovered in a Chinese tomb in 1957, and dates roughly to 100 BC. The first paper with writing on it is also of Chinese origin, and can be traced to about 110 AD. Supposedly, this paper was made by a process developed by Ts'ai Lun, the "chief eunuch" in the emperor's court. Why the emperor needed eunuchs isn't exactly clear, but to guard the ladies of the court would be a good guess. In any case, Ts'ai Lun apparently had some time on his hands, and discovered that macerating hemp fibers, old rags, and scrapings from the inner bark of mulberry trees with water, and then spreading the resulting pulp thinly on a drying frame, resulted in a material suitable for writing.

Amazingly, news of this discovery did not spread to the Western world for about 1,000 years. Europeans recorded their history on parchment, laboriously made from animal skins. When word finally reached Europe through Arabs who had learned about papermaking from the Chinese, one would have expected the Church to jump on the new technology. Such was not the case. Parchment was the only material fit to carry the Sacred Word, the Church maintained, and called papermaking a "pagan art." Initially there was not much opposition to this curious view, because papermaking was not an easy task for Europeans. There were no mulberry trees, which seemed to be the key to Chinese paper. Finally, they turned to hemp fibers, cotton, and linen rags as raw materials. These were boiled in water to a point of disintegration, and then pounded into a pulp before being poured into drying trays. Treatment with animal gelatin usually followed to prevent water absorption and to reduce the spreading of the ink. Each sheet had to be made by hand, but the paper was of remarkably good quality, as witnessed by the spectacular condition of manuscripts such as the Gutenberg Bible. (Gutenberg printed Bibles both on parchment and on paper, so his work represents the transition from the old to the new.) Soon, as more and more people learned to read, and as the Industrial Revolution began to pick up steam, rags could no longer meet the demand for paper manufacture. This forced the English to pass a law that all burial garments had to be made of wool, a substance that could not be used to make paper. By the mid-nineteenth century, the shortage was so severe that America actually imported linen wrappings from Egyptian mummies to make paper. And then came a breakthrough. Friedrich Keller, in Germany, devised a method of making paper from trees!

This idea that paper really does grow on trees had actually been brewing since the early eighteenth century. That's when René de Réamur, a French mathematician, physicist, and na-ture lover, had trouble publishing his research due to a simple lack of paper. Then, one day, while out on one of his nature walks, he happened to take a close look at a nest fashioned by North American wasps. Its light thin walls looked as if they were made of paper! Several months of study led him to the realization that the insects dined on twigs, which their digestive system somehow converted to paper. In 1719, he excitedly reported to the French Royal Academy that the American wasp makes a fine paper by extracting the fiber of common wood. "They teach us," he said, "that one can make paper from fibers of plants without the use of rags or linens, and seem to invite us to try whether we cannot make fine and good paper from the use of certain woods."

De Réamur was not an experimentalist, but Jacob Schaffer, a German clergyman, was. He successfully mimicked the work of the wasps and produced paper samples from various woods. Friedrich Keller, another German, capitalized on the idea and devised a papermaking process based on chipping wood and then beating the chips into pulp. The pulp could be mixed with water, and the resulting slurry poured through a fine screen. When dried, the residue from this "mechanical pulping" process yielded sheets of paper.

Joy, however, was short-lived, as the newfangled paper proved to be of poor quality. Chemists soon discovered why. The pulping process degraded the wood fibers into shorter fragments, which weakened the paper, and, unlike cotton or linen, wood pulp contained a substance called lignin, which caused the paper to discolor readily. As revealed by examination under a microscope, wood is made up of vertical stacks of hollow fibers, anywhere from 1 to 3 millimeters long, held together with the glue-like lignin. Because lignin is such a strong binding agent, it was difficult to separate intact fibers through mechanical grinding. The fibers ended up being ripped into smaller fragments, which give a weaker pulp than one made of the longer fibers found in cotton or linen.

And then there was the problem of yellowing. Lignin reacted with oxygen and light to produce colored molecules, which were responsible for the discoloration of the paper. So chemists went to work on trying to dissolve the lignin out of the pulp. They soon discovered that this could be done by "chemical pulping," a process that involved boiling wood chips in a sulfite solution. It was superior to mechanical pulping, but the process also degraded some of the cellulose fibers. Good enough for newsprint, but not for quality paper. Around 1880, German paper manufacturers introduced the "kraft" process. Digesting wood pulp with a mixture of sodium sulfide and sodium hydroxide yielded paper that was strong ("kraft" is German for strong), but that still yellowed because of residual lignin. It was great for many uses, including grocery bags (we still use kraft paper for these), but had to be bleached if it were to be converted into writing paper.

Bleaching was not a problem, as chemists were already familiar with the ability of chlorine to remove color from fabrics. It had the same effect on lignin still left in the paper. But unfortunately, chlorine also degraded cellulose. (Just think of what happens if you leave bleach on a cotton fabric too long.) As we later learned, it also reacted with components of lignin to produce the notorious dioxins, compounds that are toxic in minute concentrations. This eventually forced the industry to look for alternative bleaching methods. Today, most bleaching is carried out with oxygen or chlorine dioxide, which do not produce dioxins. Another problem that plagued paper manufacturers was the smell produced by the delignification process. Anyone who has ever been around a paper mill will agree that the aromas of methyl mercaptan or dimethyl sulfide do not make treasured memories. Mercifully, modern pollution control equipment has dramatically reduced the emissions.

By the late 1800s, many, but certainly not all, of the paper production problems had been solved. A major concern was cellulose's natural affinity for water. Paper lacked resistance to moisture, which meant that any ink applied would spread too easily. But chemists were not going to be stymied by this. Somehow, they had to find a way to waterproof the paper's surface. And they did. The water repellant properties of rosin, a substance that could be extracted from the southern pine tree, were well known. But how could it be applied to paper? Aluminum sulfate (alum) was already used at the time as a "mordant," a substance that allowed dyes to stick to fabrics, so applying the same chemistry to paper was logical. It worked. This was the very first example of "sizing" paper. The word derives from the Latin "assidere," to set in place. Basically, waterproofing chemicals are set in place on the surface of the cellulosic fibers.

Paper was further improved by the addition of materials such as starch and kaolin (a type of clay), which filled in some of the pores between the fibers, and titanium dioxide, which added opacity and brightness. Beautiful printed pages began to roll off the presses. Everything seemed hunky-dory. But this didn't last long. Aluminum sulfate, you see, is an acidic substance. And acids break the glucose-glucose linkages in cellulose. This weakens the paper, and discolors it to boot. The fragments of cellulose now can be oxidized by air to molecules that contain "aldehyde" groupings, and such fragments are yellow. That's why books printed on acid paper begin to turn yellow and crumble after thirty years or so, even if the lignin has been removed. That's why the death knell now sounds for millions of books and manuscripts stored in libraries around the world.

Intense efforts have been mounted to save these works. Deacidification processes, ranging from rinsing individual sheets in alkaline solutions (calcium hydroxide, for example) to exposing whole books to gaseous bases such as diethyl zinc, have met with various degrees of success, but cannot possibly be applied to millions and millions of aging books. As with many other things in life, prevention is better than treatment. And once again, chemists have taken up the challenge. Sizing materials that do not leave an acid residue have been developed. They have names like "alkyl ketene dimers" or "alkyl succinic anhydrides," which do not roll easily off the tongue. But water does roll easily off paper treated with them. And instead of an acid residue, these compounds leave an alkaline one. That also means that the expensive titanium dioxide whitener can be replaced by cheaper calcium carbonate. This cannot be used in acid papers because it reacts with acids to liberate carbon dioxide gas. Alkaline paper that uses this technology was introduced around 1990, and is taking over from acid papers. Tests show not only that alkaline paper is stronger, is more readily recycled, and can last for hundreds of years, but that its manufacture is less polluting, requires less energy, and leads to less machine corrosion.

There have also been dramatic developments in mechanical pulping. It turns out that if wood chips are heated with steam, the lignin softens enough so that the wood fibers can be pulled apart without much damage. The lignin is left in, but can be decolorized with hydrogen peroxide. Whereas, twenty-five years ago, mechanical pulp had to be blended with bleached kraft pulp to make paper strong enough for newsprint, today it can often be made with 100 percent "thermomechanical" pulp. Paper is clearly constantly being improved. So save this book and read it again in 100 years. You'll be amazed by how some of the problems described here will have been solved.

THE BIRTH OF THE PILL

Russell Marker had no interest in contraception. To his dying day, he was uncomfortable with the title "Father of the Birth Control Pill," a name that numerous articles singing the praises of his chemical exploits bestowed upon him. But the truth is that the pioneering work of this remarkable but peculiar man did indeed lead to the development of "The Pill." Actually, it did more than that. Marker's ingenious chemistry gave birth to the synthetic steroid industry, legal and illegal.

Marker graduated with a master's degree in chemistry from the University of Maryland in 1924 and immediately enrolled in a PhD program. He was so adept at laboratory work that within a year, he had amassed enough data to meet the requirements for a degree. But there was a slight problem. Doctoral candidates were obliged to take courses as well as carry out research. Marker had no desire to take these, particularly the physical chemistry courses that he considered irrelevant to his work, and therefore a waste of time. His research director, Professor Morris Kharasch, warned him that if he refused to take the prescribed courses he would not be granted a degree, and he would end up as a "urine analyst." At that point, the stubborn Marker left the university and found a job at the Ethyl Corporation, where he developed the octane rating system for gasoline still used today. That alone would have secured his fame in chemical history, but Marker sought new horizons. He became interested in the fledgling field of steroid chemistry, and secured a research position funded by the Parke-Davis pharmaceutical company at Penn State College.

Steroids are a class of naturally occurring compounds that have a common basic molecular structure consisting of four rings of carbon atoms. The female hormones progesterone and estradiol are typical examples, and indeed were some of the first steroids ever identified. But they were difficult to come by. In the early 1930s, it took 4 tons of sow ovaries to isolate 12 milligrams of estradiol, and dozens of pregnant sows were needed to produce a few milligrams of progesterone. As the name implies ("for gestation"), progesterone was found to be the hormone that prepared the uterus for the implantation of a fertilized egg. This raised the possibility that the compound could be used to prevent miscarriage and to treat certain menstrual disorders. There was even a theory that progesterone could be useful in the treatment of cervical cancer. Extraction from natural sources was very difficult, and the only available progesterone came from a multi-step synthesis starting from cholesterol, the first steroid ever isolated, way back in the 1700s. It sold for about $80 a gram, a stunning price in those days. Could it possibly be produced more cheaply? Now that was the kind of challenge that Russell Marker liked.

Marker knew that steroids could also be found in plant products. He was particularly interested in sarsaponegin, which had been isolated from the sarsaparilla root in 1914 and had a chemical similarity to progesterone. By 1938 the clever chemist had developed a series of reactions to convert sarsaponegin into progesterone, but the starting material was still hard to come by. Could another plant be a better source of steroids? Marker began to look through botany texts for plants that looked like ones known to produce steroids and was taken by pictures of the Dioscoria species. The scientific literature revealed that a steroid known as diosgenin had indeed been extracted from these plants. He could probably convert diosgenin into progesterone, Marker thought, and began a systematic search for plants of the genus Dioscoria, hoping to find a variety that produced diosgenin in good yield. He first scoured the southwest US, but eventually ended up in Mexico, and it was there that his efforts finally bore fruit. Actually, they bore a root. The root of a Mexican yam looked like a good candidate to Marker, and he smuggled a sample back into the US, where analysis revealed that it did indeed contain a wealth of diosgenin. It took Marker just five chemical steps to convert diosgenin to progesterone, and eight to make testosterone, the main male hormone.

Surprisingly, Marker was unable to interest major pharmaceutical companies in his synthesis, so he decided to go it alone. He left his wife, moved to Mexico, and set up a crude laboratory to extract diosgenin from yams and convert it to progesterone. Marker single-handedly gathered about 10 tons of root, from which he produced an astounding 2 kilograms of progesterone. Its commercial value was about $160,000, but Marker had no idea how to market his windfall. Looking through the Mexico City telephone book, he chanced upon "Laboratorios Hormona," which sounded like a potentially interesting connection. Marker took his progesterone, wrapped it in newspaper, and marched down to meet Emeric Somlo and Frederick Lehmann, the two European refugees who had founded the small drug-marketing firm. Never before had anyone seen so much progesterone at one time! Somlo and Lehmann were duly impressed, and before long, the three had formed a new company, Syntex, to begin commercial production of progesterone. Within a year they had produced 30 kilograms of progesterone, and were supplying various pharmaceutical companies, which sold the hormone to treat menstrual and menopausal problems.

It didn't take long for the mercurial Marker to have a falling-out with his partners, and in 1945, he left Syntex in a huff, taking many of the details of progesterone synthesis with him. His former partners, Emeric Somlo and Frederick Lehmann, were desperate to find a chemist who could pick up the pieces and realize the profits that synthetic progesterone promised. Physicians and their patients were excited about the substance that offered hope for preventing miscarriage and for treating the symptoms of menopause. Somlo and Lehmann searched high and low, and finally found their man in Cuba.

George Rosenkrantz was born in Hungary and trained in Switzerland under Nobel Prize winner Leopold Ruzicka, who had identified the sex hormones as belonging to the family of compounds known as steroids. In the 1930s, Ruzicka offered positions to many Jewish scientists who fled Eastern Europe, but he became particularly closely associated with Rosenkrantz. Although Switzerland was nominally neutral, Rosenkrantz worried that his mentor's friendship with Jews might affect his career, and therefore decided to head west, and eventually ended up in Cuba. Here he maintained his interest in steroids, and having heard about Marker's work, even produced small amounts of progesterone from sarsaparilla root. Somlo and Lehmann heard about this and invited Rosenkrantz for an interview. When in an impromptu performance he carried out one of the steps in the progesterone synthesis, the job was his. Within two months, Rosenkrantz had resuscitated Syntex, and the company began to fulfill its orders for progesterone. He then went on to synthesize testosterone, the main male sex hormone, using progesterone as starting material, and managed to convert it to the female hormone estrone. "Adam goes into the test tube," Rosenkrantz quipped, "and Eve comes out."

Then Rosenkrantz set his sights on a target that he thought would be even more lucrative than progesterone. In 1948, a researcher at the Mayo Clinic in Minnesota tantalized the medical community with a film he had made of bedridden arthritic patients who got up and danced after being treated with an apparently miraculous substance. The drug was cortisone, an adrenal gland hormone, which had previously shown potential in the treatment of various diseases, including arthritis, asthma, skin disorders, and even leukemia. But extraction from adrenal glands was extremely difficult, and the only viable source was a complicated thirty-six-step synthesis that had been developed by the Merck Company. Now the dancing patients triggered a race between Merck, a group at Harvard, and Rosenkrantz's team to come up with a practical synthesis. At Syntex, the task fell to a promising young chemist, Carl

Djerassi, whom Rosenkrantz had enticed away from the CIBA pharmaceutical company in New Jersey. In just two years, *Life* magazine trumpeted Syntex's triumph with the headline "Cortisone from a Giant Yam: Scientists with Average Age of 27 Find Big Supply in Mexican Root."

Djerassi's cortisone synthesis did not prove to be an economic success because, just a few months later, the Upjohn Company came out with a cheap process to make hydrocortisone, a compound that basically performed as well as cortisone. Syntex, however, was not left out in the cold. The raw material that Upjohn needed to make hydrocortisone was progesterone, which could only be supplied in the desired amounts by Syntex! Now the company had funds to pursue further research. One of the synthetic targets was estradiol, the female hormone that was being explored for the treatment of menopausal symptoms. Rosenkrantz and Djerassi thought that estradiol could perhaps be made from progesterone, of which of course they had a plentiful supply.

The synthesis proved unsuccessful, but one of the compounds that emerged, 19-norprogesterone, aroused their attention. When tested, it performed like progesterone, but was far more potent. Now, this was interesting! One of the drawbacks of progesterone had been the need to inject it into the bloodstream, because it was not stable to stomach acid. But perhaps enough of the more potent compound would survive in the stomach, allowing oral use. Alas, such was not the case. Djerassi's appetite was whetted, though, and he searched through the scientific literature for steroids that were stable to acid. He discovered that a German chemist, Hans Inhoffen, had found a way to alter the estradiol molecule to make it stable to acid. Djerassi now applied this reaction to his 19-norprogesterone and produced norethindrone, which behaved just like progesterone, but was more potent and, amazingly, could be taken orally! This would be a great improvement. Women did not relish progesterone injections even if these were effective treatments for fertility and menstrual problems. Little did Djerassi dream that his compound would eventually be referred to as the "first oral contraceptive ever synthesized." Neither did he dream that Frank Colton at Searle would soon come up with a similar substance, norethynodrel, which would beat his compound to be the active ingredient in the first birth control pill.

The connection to birth control was made by Gregory Pincus, a biologist, who had been approached by birth control activists Margaret Sanger and Katherine McCormick about the possibility of coming up with a physiological way to prevent pregnancy. Pincus had made somewhat of a name for himself by fertilizing rabbit eggs in a test tube (the press sometimes depicted him as a "Dr. Frankenstein") and took on the challenge. Knowing that a pregnant woman could not get pregnant a second time because ovulation was suppressed by progesterone, Pincus immediately recognized the potential of Djerassi's and Colton's work. He enlisted the help of John Rock, a Harvard fertility expert who, in his practice, had witnessed the suffering of women burdened by unwanted pregnancies. Tests carried out on volunteers in Puerto Rico in 1956 showed that norethynodrel was effective in preventing pregnancy. Pincus had used Searle's compound because the company had previously supported its research. By 1957 both norethynodrel and norethindrone had been approved as drugs, but only for menstrual and fertility disorders.

There was great hesitancy among pharmaceutical companies to apply for approval of a contraceptive pill because of opposition from the Catholic Church and other religious groups. Searle finally took the bold step and in 1960 received FDA approval for its Enovid as the first birth control pill. It contained 10 milligrams of norethynodrel and a little estrogen to reduce side effects. Syntex had licensed its drug to Parke-Davis, but the company was worried that religious opposition would lead to a boycott of all its products if it got into the chemical contraception game. Thus, Djerassi's norethindrone, the "first oral contraceptive ever synthesized," was not the active ingredient in the first commercial birth control pill. In 1962, Djerassi's compound finally made it to market under agreement with Ortho pharmaceuticals. Syntex went on to become a multibillion-dollar company, eventually introducing Naprosyn and Anaprox, two widely sold non-steroidal anti-inflammatory drugs (NSAIDs). But of course none of that would have happened without Russell Marker's original work on steroids. So who was the father of "The Pill," the drug that, for more than forty-five years, more people have taken than any other prescribed medicine in the world? Well, you decide.

THE GREATEST INVENTOR

He holds 1,093 patents. His inventions ranged from an electric voting machine and the stock ticker to motion picture cameras and the phonograph. He produced the world's first feature film and first storage battery, perfected kilns to make superior cement, and developed a process for extracting iron ore from the ground. Not bad for someone with only a few months of formal education. And that is all young Thomas Edison had! His mother, infuriated when her son was labeled as "addled" by his first teacher, had resolved to educate the boy at home. Mrs. Edison did a wonderful job, nurturing his love of learning, introducing him to the great works of literature, and encouraging him to study those things that interested him the most. As the man who grew to be perhaps the world's greatest inventor later said, "My mother was the making of me. She understood me; she let me follow my bent."

Most people don't know that young Tom Edison's first "bent" was chemistry! His passion was triggered by R. G. Parker's book *School of Natural Philosophy,* which his mother gave him in 1857, when Edison was ten years old. The book described various experiments that could be done at home, and before long, Tom had built a chemistry lab in his basement. This is where he conducted his legendary experiment with friend Michael Oates. Edison thought that if his friend drank a mixture that would produce a gas in his stomach, he would be able to fly like a gas-filled balloon. Michael, gas and all, stayed put, and got sick. This put a crimp into Tom's basement chemical investigations for a while, but soon he was back at it, albeit in a different location. On a train! At the age of twelve, young Edison got a job as a newsboy on the train that ran daily from Port Huron to Detroit, but didn't know what to do during a five-hour layover before the train returned. He had managed to get permission to move the lab from his cellar aboard the baggage car so that he could continue his experiments. Everything was fine until a piece of white phosphorus burst into flame and set the baggage car on fire. The conductor had had enough of Tom's foolery, and threw him and his chemicals off the train. In spite of this misadventure, Edison never lost his enthusiasm for chemistry, although he did become enchanted with the mechanical and electrical devices that would make his fame and fortune. But even here his chemical interest would come in handy!

Edison's favorite and most original invention was the phonograph. The earliest version used a needle that moved in response to sound to etch a pattern into tin foil that covered a rotating drum. To play back the recording, the process was reversed. The needle was connected to a diaphragm that vibrated as the needle traced the pattern in the foil, causing the air around the diaphragm to move. Our ears perceive such rhythmic pulsations of air as sound. Edison's first tinny recording of "Mary Had a Little Lamb," in 1877, was an astounding breakthrough, but foil was very fragile, and early recordings could be played only a couple of times. A more reliable material was needed, so Edison went back to his chemical roots. Using stearin, the rendered fat of cud-chewing animals, he concocted a wax base of stearic acid, sodium stearate, and aluminum stearate. To achieve the right consistency, he experimented with mixing in various naturally occurring waxes, like carnauba, whale wax, beeswax, or ceresin wax. The components had to be blended together in a heated vat—a dangerous business. One day, in 1899, a batch exploded, leaving Edison swathed in bandages for weeks. He did eventually come up with a mix to manufacture cylinders on which sound could be more permanently recorded. By then Edison had set up his "invention factory" in Menlo Park, New Jersey, where he and his crew of insomniacs doggedly worked on inventions. Modifications such as adding lead and asphalt to the wax yielded better recordings. Edison never gave up trying to improve his cylinders, and eventually switched to the novel plastics celluloid and bakelite.

Edison's chemical ingenuity was also evident in his electrical endeavors. He had designed a system to distribute electricity, but there was the question of how to charge for the service. Edison was a businessman, reportedly often saying that if an invention could not be sold, he was not interested in it. He knew all about the process of electrolysis, whereby the passage of electricity through a solution can cause substances from solution to plate out on an electrode. So he designed an electric meter that consisted of two copper plates dipped into a solution of copper sulfate. When direct current passed through the solution, one of the copper plates served as the positive electrode, the other as the negative. This meant that positive copper ions from solution were attracted to the negative electrode (cathode), and plated out. At the same time, copper from the positive electrode (anode) dissolved in the solution. The net result was a change in the weight of the plates in proportion to the amount of electricity that passed through the solution.

Meter readers would then periodically come, switch off the power, and remove the copper plates, replacing them with new ones. The used plates were taken away and weighed. On the basis of the change in weight, the customer would be charged for the electricity used. The electric meters were often located in unheated basements, and there was a chance that the solution would freeze. Edison solved the problem neatly. He incorporated a lightbulb into the meter, which produced enough heat to warm up the solution. The bulbs, naturally, were to be bought from Edison's company. Edison was a clever man.

Pepper's Ghost

I think it must have been around 1960 when I was attacked by a gorilla. It happened at Belmont Park, a classic old-time amusement park in Montreal, where sideshows were a hot attraction. Large canvases with peeling paint advertised fat ladies, sword swallowers, living skeletons, and a "gorgeous woman who would magically transform into a ferocious gorilla in front of the spectator's eyes." This I had to see!

After paying a quarter, I was ushered to a corner of a large tent where a cabinet, somewhat larger than a phone booth, had been set up. Judging by the gaudy picture of a half woman–half ape that adorned the side of the contraption, it was clear that this was where the miraculous transformation would take place. Indeed, as the lights came on, we saw a lady shackled to two posts inside the cabinet. She seemed pretty harmless to me, so why she had to be shackled wasn't clear. I thought the whole thing would probably turn out to be some sort of scam, like the "living skeleton," who was actually just a very thin man. But no! Truly as if by magic, the lady slowly morphed into a gorilla that growled fiercely and began to vigorously rattle its chains. Then, as the noise got louder, and as the gorilla's struggle intensified, the shackles suddenly gave way, and the beast bolted from the cabinet, scaring the daylights out of all of us. We scampered toward the exit. When I looked back, the man in the monkey suit was gone, and the next group was already being assembled in front of the cabinet.

I didn't realize it at the time, but I had just witnessed one of the greatest scientific illusions ever devised. I had just seen "Pepper's Ghost." Actually, if you really want to get technical, it was really "Dirck's Ghost," as modified by Pepper. To understand what all of this means, come back with me for a moment to Victorian London. The date is December 24, 1862, and you're sitting in the audience at The Royal Polytechnic Institute in Regent Street. You have heard of the wonderful lectures put on by "Professor" John Henry Pepper, and have come to be entertained by his scientific experiments. The Polytechnic, founded in 1838, was sort of a permanent science fair to which the public flocked to view the latest technological marvels and listen to fascinating lectures. The most entertaining presentations were by John Pepper, who had been captivated by chemistry since childhood and now reveled in bringing science to the public in a theatrical fashion. He explained the workings of steam engines, diving bells, and poisons. Then, using a projection microscope and a giant screen, he terrified the audience by showing microbes cavorting in a drop of London drinking water.

On this December day, though, the professor had something special for his audience. The curtain went up to reveal a scene from Charles Dickens' *The Haunted Man*. A student was seen hunched over a desk, when suddenly, a ghostly skeleton appeared and seemed to float right through him. The audience burst into spontaneous applause at the appearance of "Pepper's Ghost."

Pepper, as manager of the Polytechnic, had always been on the lookout for novel acts and demonstrations. So when a Liverpool civil engineer named Henry Dircks approached him with an invention he claimed would astound audiences, Pepper was ready to listen. Dircks asked Pepper to look down into a box he was holding. As Dircks manipulated a flap on the side of the box, Pepper was absolutely flabbergasted to see the appearance of little ghostly characters. The "ghosts," Dircks explained, were reflections of figures hidden in the box in front of a glass plate. Unless they were lit, all that was visible was the back of the box through the glass plate. But when a flap was opened to let in some light, the figures were reflected in the glass, with their image appearing to be as far behind the glass as the figures were in front of it.

Actually, we have all witnessed this phenomenon. Just think about what you see when you look out through a window from a lit room into the dark night. Reflections of objects in the room appear to be floating outside. To his credit, Pepper recognized the potential of Dircks' discovery for producing theatrical effects. He designed a large glass plate for the stage, tilted toward the audience at a 45-degree angle. A second, essentially identical stage was built where the orchestra pit might normally be located. This is where the "ghosts" cavorted, in front of black drapes. When this stage was dark, the audience just saw the scene behind the glass plate. But as the limelight slowly came on, and the illumination on the real stage dimmed, the reflections of the characters appeared as if they were transparent ghosts. With proper synchronization, they could even interact with the actors on stage.

Some in the audience believed they had seen a real paranormal event, but Pepper was quick to explain that it was all a trick of science. He then went on to castigate the spiritualist mediums, who were very popular at the time, suggesting that they also used tricks to prey upon the gullible. Today, if you want to experience the most spectacular display of Pepper's Ghost ever created, just visit the Haunted House at a Disney theme park. Or, if you're lucky, you can still find a carnival where a woman turns into a gorilla.

STRADIVARIUS OR NAGYVARYUS?

What do you think the leading violinists in the world will be playing 300 years from now? A Stradivarius? A Hutchinsius? A Nagyvaryus? Or perhaps the unthinkable: a syntheticus? Let me tell you, I have dealt with a number of controversial issues in my career, but the vocal battles between the critics and supporters of pesticides, genetically modified foods, or artificial sweeteners are like playground chatter when compared to the mudslinging and sniping that take place between violin makers with different theories about how to produce the best instrument. Basically, it comes down to an emotionally charged three-way fight between physics, chemistry, and tradition.

On at least one count, there is agreement among the camps. The Holy Grail for luthiers is to be found in matching the superb quality of instruments made in the eighteenth century in Cremona, Italy, by master craftsmen like Antonio Stradivari and Giuseppe Guarneri. Only a few hundred of these remain in existence, and many are valued in the millions of dollars.

Playing a Stradivarius has been described as a holy experience, one unmatched by playing a modern instrument. But now the secret of the Strad may have been finally discovered. The question is, by whom?

Carleen Maley Hutchins retired from teaching high school science to devote her life to the grand old art of violin making. Actually, she had in mind to look at the science behind the art. For some twenty years, she collaborated with Harvard physicist Frederick Saunders to study the vibrations generated in the sound box of the violin when its strings were set to oscillate. Drawing a bow across the strings makes them vibrate, setting the surrounding air into motion. The moving air in turn causes the panels of the violin to vibrate, producing sound. Hutchins used an electric tone generator and studied the way Christmas glitter vibrated when sprinkled on the top and back plates used to make the sound box. She concluded that the key to producing the most pleasing tones was the mass and thickness of the wood and the exact placement of the "bass bar" and "sound post" inside the box. Still, that wasn't all. According to Hutchins, the more a violin is played, the better it sounds. She claims that decades worth of vibrations alter the structure of the wood, improving its resonant qualities. That's why Hutchins attempts to give her instruments a head start by exposing the wood to some 1,500 hours of classical music before she puts her violins on the market. In a hundred years or so, she says, they should sound just like a Stradivarius!

"Humbug!" retorts Joseph Nagyvary, a former professor of biochemistry at Texas A&M University. The secret of the Stradivarius lies not in the physics of the sound box, but in the chemistry of the wood and varnish. Of course, the construction of the sound box is important, he admits, but it is not the critical feature, given that the classic violins have been analyzed and copied down to fractions of millimeters without their magnificent sound being reproduced. At least not until Nagyvary got into the game. Stimulated once being given the chance to play the same violin that Einstein had played, the chemist attacked the mystery of the Stradivarius by subjecting a few fragments from violins of the era to scanning electron micrography and x-ray spectroscopy. He found remnants of fungi in the wood, and suggests this can be traced to its having been soaked in seawater for a long period, probably because in Stradivarius' time, logs used to be floated downriver to the Adriatic Sea. This changed the properties of the wood, and resulted in absorption of minerals from the water. Nagyvary also hypothesized that boron and aluminum present in the wood may have come from borax and alum used to keep the wood from rotting. He also found evidence that Stradivari used a complex varnish composed of tree bark exudates, perhaps guar gum, mixed with finely ground glass and other mineral powders. According to Nagyvary, credit for the spectacular sound of the Stradivarius should therefore go to the unknown chemist who provided the preservatives and varnish.

Professor Nagyvary has experimented for over thirty years with various formulas, and is now convinced that he has essentially reproduced the magic. Some of his violins have sold, like those of Hutchins, for as much as $15,000. As a result, he says he has received hate mail from violin dealers and manufacturers who feel threatened, and who bristle at the suggestion that the great Stradivarius may not have realized what made his instruments great. Hutchins has also been derided by the traditionalists, who just don't want science injected into their art. So you can imagine what they have to say about a "Maccaferrius."

Mario Maccaferri was a traditional maker of guitars and violins, at least until he visited the New York World's Fair in 1939. There he was captivated by a display of plastics, and after the war, he managed to get his hands on some polystyrene injection molding equipment. He made a small fortune by making plastic clothespins, and then began a foray into plastic instruments by making a ukelele that was soon made famous by the entertainer Arthur Godfrey on his television programs. Millions of these were sold, and were followed by plastic guitars and violins. The violins were not a huge success among elite players because they did not compare to traditional instruments in sound quality. But a principle had been introduced. Some experts today claim that plastic's ability to be molded to exact specifications will eventually produce an outstanding instrument.

In the meantime, Nagyvary seems to be winning the battle. In a violin duel staged at Texas A&M, a world-class violinist played a Nagyvaryus, and then a Stradivarius, behind a screen. Both the invited experts and the audience rated Nagyvary's violin slightly higher. For now, at least, chemistry has triumphed! But how will it sound in 300 years?

It's Dynamite!

Alfred Nobel wasn't in the best of health, but he knew he wasn't dead. Yet there was his obituary, prominently featured in the morning newspaper. To make matters worse, not only had the newspaper killed him off prematurely, it had described him as a man who "became rich by finding a way to kill more people faster than ever before." The French press service that provided the story had made a mistake. It was actually Alfred's older brother Ludvig who had died, while vacationing in Cannes, but a reporter had gotten the brothers mixed up. Alfred was deeply disturbed by this chance preview of how the world would remember him. Yes, he had invented dynamite and gelignite, the most powerful explosives known at the time, but he had always envisaged that they would be used to benefit mankind. Indeed, he had spoken of producing a substance of "such frightful efficacy for wholesale destruction that it would make wars impossible." Unfortunately, he was wrong.

Nobel was born in Sweden, but spent his early years in St. Petersburg in Russia, where his inventor father had set up a small business developing sea mines for the Russian government. Young Alfred had ambitions of becoming a writer, but his father thought that a scientific career would be more practical. So he sent sixteen-year-old Alfred to apprentice in the laboratory of the noted French chemist Theophile Pelouze. It was here that he met Ascanio Sobrero, an Italian chemist, who told him about a fascinating substance he had discovered. "Pyroglycerine," Nobel learned, was an oily liquid that exploded with great vigor when detonated. Sobrero had made it by reacting a mixture of nitric and sulfuric acids with glycerine, a substance readily available by treating fats with sodium hydroxide. He had gotten the idea from a story about a chance discovery made in 1838 by Friedrich Schonbein, a professor of chemistry at the University of Basel in Switzerland. Schonbein, as the story goes, was experimenting in his kitchen with a mixture of nitric and sulfuric acids, which he accidentally spilled. He quickly picked up his wife's cotton apron and wiped up the mess. When Schonbein tried to dry the apron by hanging it near a stove, it burst into flame and disappeared in a flash. He realized that cellulose, the basic component of cotton, had somehow reacted with the acids to create an explosive material.

Sobrero realized that glycerol and cellulose shared some chemical features, and he wondered what would happen if he reacted it with the mix of acids that Schonbein had used. The results were remarkable. The nitric acid converted glycerine into Sobrero's "pyroglycerine," which in chemical lingo was better described as "nitroglycerine." When heated, it just burned. But as the temperature reached 220°C (428°F), nitroglycerine exploded, although not always in a predictable fashion. The yellow liquid was also sensitive to shock, and it seemed to Nobel that if nitroglycerine were to be used as an explosive, a reliable detonation system would have to be found.

Alfred suggested to his father that they focus their attention on making nitroglycerine on a large scale. Immanuel Nobel did not need much convincing because his factory in St. Petersburg, which had been very profitable during the Crimean War, now faced bankruptcy. The family moved back to Sweden and set up a factory to produce nitroglycerine. Tragedy struck almost immediately, when an explosion killed Emil, the youngest son. The nitration of glycerine was a dangerous business. So dangerous that in some cases, the workers who monitored the reaction were made to sit on one-legged stools so that they would immediately wake up should they doze off. One would think, though, that sitting in front of a bubbling kettle frothing with brown fumes of nitrogen oxides, containing the most powerful explosive known to mankind, would have been ample motivation to stay awake.

Making nitroglycerine wasn't the only problem. An even bigger concern was how to detonate it. Alfred solved this problem with his invention of the mercury fulminate blasting cap. But without a doubt, Nobel's greatest contribution was the invention of dynamite, which safely harnessed the energy of nitroglycerine. He had long considered the idea of mixing nitroglycerine with some solid material with the hope of decreasing its shock sensitivity. Finally, Nobel hit on a type of silica known as diatomaceous earth, which was ideal. The sticks of dynamite could be safely transported and would only explode when triggered with a blasting cap. Dynamite would change the world. It would allow the Panama Canal to be built, but, contrary to Nobel's hopes, it would also take warfare to a new level.

Alfred Nobel had loathed war all his life and was stunned when his obituary referred to him as a "merchant of death." He vowed that he would not be remembered as such! So he decided to leave his immense fortune to foster science, literature, and peace. The Nobel Prizes were born! And it was all because a journalist did not check his facts.

THE POX—BOTH COW AND SMALL

The inscription on the statue in Kensington Gardens, London, simply says "Jenner." The sculptor obviously thought further details were unnecessary since, after all, just about everyone has heard the story of the English country doctor who discovered a way of protecting people against that most dreaded of diseases, smallpox. Well, as is so often the case with historical accounts, the story that people generally hear is incomplete. Edward Jenner deserves credit for his tireless effort to promote vaccination, but he was certainly not the first person to come up with the idea of inoculating people against smallpox with material taken from pustules on the skin of cowpox victims. An English gentleman farmer, Benjamin Jesty, inoculated his family with cowpox extract some twenty years before Jenner's "discovery"!

Smallpox is a horrific disease that kills about 30 percent of its victims, leaving the rest scarred and often blind. As recently as the past century, it was responsible for more deaths than all wars combined. So it comes as no surprise that, throughout history, numerous attempts have been made to ward off this scourge. And one method, introduced in China in the tenth century, actually worked. That is, when it didn't kill the recipient. Powdered scab taken from smallpox pustules was blown up the nose of people who desired to be protected against the disease. Those who didn't come down with smallpox after this procedure were indeed protected for life. Word spread westward, and modifications to this process were introduced. Lady Mary Wortley Montagu, the wife of Britain's ambassador to Turkey in the eighteenth century, described how old women in Constantinople would scratch a vein in children and introduce an extract taken from someone who had suffered from smallpox. This made sense because, even at the time, it was understood that if someone survived smallpox, they would never get the disease again. So why not give the disease to the young and healthy who had the best chance of recovery? Indeed, by the early 1700s, healthy people in England were encouraged to be inoculated with matter taken from a patient sick with a mild attack of smallpox, but many were obviously reticent. The inoculation idea crossed the ocean as well. In the late 1760s, James Latham, a surgeon with the British military in Quebec, was inoculating both soldiers and civilians with the pox extract.

In 1774, an epidemic of smallpox struck the village of Yetminster, where Benjamin Jesty was a prosperous farmer. Knowing that about one in fifty people died from smallpox "variolation," as the technique was known, he was unwilling to put his family at risk. He felt confident that a better form of protection against smallpox was available, and he thought he knew just what it was! Jesty, like many others at the time, had heard stories of people who avoided smallpox because they had previously been afflicted with a much milder disease, known as cowpox. This disease was endemic in some herds and could be transferred to people, especially milkmaids. Jesty himself had two milkmaids in his employ who, after having had cowpox, nursed relatives with smallpox without getting the disease. Impressed by this, the farmer took his wife and two sons out to a field where a herd of cows with symptoms of cowpox was grazing. Using a knitting needle, he scratched a lesion on a cow's udder and transferred the material to a small incision he made on the elbow of his wife and sons. The three remained free of smallpox for the rest of their lives!

It is impossible to say whether or not Edward Jenner knew of Jesty's ingenious idea, but being a physician, he was certainly aware of the inoculation process with live smallpox material. In fact it was his use of variolation that allowed Jenner to make his historic observation: some patients developed no reaction to the inoculation at all! When he questioned them, he learned that they had all previously had cowpox. That's probably when Jenner recalled the words of his mentor, famed London surgeon Dr. John Hunter. "Why think? Why not try the experiment?" And so he did.

In 1796, an epidemic of cowpox broke out in Jenner's village of Berkeley, and a young maid, Sarah Nelmes, consulted the doctor. She had fresh cowpox pustules on her hands, just the thing Jenner had been looking for. Amazingly, he obtained permission from the parents of eight-year-old James Phipps to try a risky procedure. He removed pus from a pustule on Sarah's arm and injected it into the boy. James developed cowpox within a week, but recovered readily. Then Jenner inoculated Phipps repeatedly with pus from a smallpox patient and found the boy to be completely protected against the disease. He followed this experiment with several others and submitted a paper about his findings to the Royal Society of England. The manuscript was rejected because it was judged to be "incredible" and "in variance with established knowledge." This forced Jenner to publish his results at his own expense in a small booklet in 1798. Within a year, vaccination, as the process was called, deriving from the Latin word for cow, was used on a wide basis, and in 1802, the British Parliament voted Jenner a grant of £10,000 in recognition of his "discovery."

Benjamin Jesty was not a scientist and never sought publicity. Indeed, it was only when Jenner received his grant that the Reverend Andrew Bell, who knew about Jesty's classic experiment, began to preach about the man "whose discovery of the efficacy of the cow pock against smallpox is so often forgotten by those who have heard of Dr. Jenner." Without a doubt it was Jenner's tireless promotion of vaccination and his numerous publications and letters to authorities that launched the massive inoculation procedures that led to one of humankind's greatest achievements, the eradication of smallpox from the world. Had Jesty published his work, he would probably be sitting next to Jenner today in Kensington Gardens.

TIN PLAGUE

Thank goodness for collectors of Civil War memorabilia who advertise their wares on the Internet. Without them I would have had a hard time finding authentic tin buttons. And I absolutely needed these to carry out my research into Napoleon's problems with the Russians and the Russian winters. After all, there is a burning question out there that has never been successfully answered. Were Napoleon's soldiers able to keep their pants on, or were they not?

Here's the often-repeated story. Napoleon had his problems with the Russian winter, particularly the winter of 1812, which was especially cold. His troops were already on the verge of defeat when an unusual bit of chemistry dealt the final blow. The buttons on the soldiers' uniforms it seems were made of tin, a metal that exhibits an interesting property at low temperatures. It disintegrates! That's because tin, like several other elements, can exist in more than one form. These "allotropes" (from the Greek meaning "other way") can have dramatically different properties. Perhaps the most familiar examples are diamond and graphite, both of which are composed only of carbon, yet are decidedly distinct substances due to the different internal arrangement of the carbon atoms. Similarly, tin atoms can be packed together in two ways. Above 13.2°C (55.76°F) we have an arrangement that is characteristic of metals, and we have the shiny, malleable material recognized as tin. But as the temperature drops, the atoms rearrange, and the metallic tin slowly changes into a non-metallic gray powder. When this happens the metal is said to suffer from "tin disease" or "tin pest."

There is no question that this is more than just a curious theoretical possibility. Just ask churchgoers in northern Europe, who have had to endure more and more false notes produced by church organs over the years. Tin has been favored as the ideal metal for organ pipes because of the appealing sounds it can produce when it vibrates. But in cold cathedrals over many years, the metal can slowly change into its crumbly non-metallic form. This of course changes the sound and in a few cases has resulted in complete destruction of the pipes. Before the chemistry of this allotropic conversion was understood, the destruction of organ pipes was attributed to be the work of the devil, who was doing his best to undermine devotion to God.

The fact, however, is that the devil never carried out his work over one winter. It took many years until the organs began to suffer from "tin plague." That's why the Napoleon story needs to be scientifically investigated. There is no doubt that disintegration of buttons would have made the waging of war a difficult venture. After all, it's hard to fight, or indeed retreat, with your pants around your ankles. But could this have happened? I tried to answer this question once and for all. The tin buttons I acquired were put to task. A couple sat in my freezer, two others took up residence in the back of the fridge, and a few braved the Montreal winter outdoors.

The fridge is actually a very appropriate place to study the behavior of tin because, before widespread refrigeration, the tin can was an ideal way to preserve food. And while Napoleon may have let his soldiers (and their pants) down with tin buttons, he helped the whole world with his determination to find a method to preserve food for his armies. This determination eventually led to the tin can. Napoleon, as many other generals before him, had discovered that soldiers do not fight well on empty stomachs. And stomachs were often empty due to the difficulty of supplying food to massive traveling armies. What food there was was often spoiled or of poor quality. So the emperor offered a prize of 12,000 francs, a healthy amount of money at the time, to anyone who could come up with a viable method of preserving food.

Nicholas Appert took up Napoleon's challenge. The son of an innkeeper, Appert had learned about brewing and pickling. He knew that these "fermentation" processes could be halted by heating, and began to wonder whether food spoilage could also be stopped in this fashion. After all, it was clear that cooked food kept longer than fresh food, although eventually it too would spoil. Years of experimentation led Appert to a critical discovery. If the food were sealed in a glass jar and then heated, it would keep for a remarkably long time. Long enough to please Napoleon, who awarded the prize to Appert in 1809. The method clearly worked, although nobody at the time understood why. Bacteria were not identified as the cause of food spoilage until another famous Frenchman, Louis Pasteur, came along later in the century.

Appert's invention came to the attention of Peter Durand, in England, who was troubled by the use of glass jars that often broke. There had to be a better way! Why not a metal container? Iron was cheap and was the first choice. But it corroded, especially when exposed to acidic foods. A coating that would protect it from the air and contents had to be found. Tin, concluded Durand, would do the job! The metal had been known since antiquity and could be easily melted and applied as a coating to iron to make tin plate. And most important, tin did not corrode. By 1818, the British Company, Donkin and Hall, was mass-producing food in tin cans. When Admiral Parry sailed to the Arctic Circle in 1824, he sustained his crew on canned food. One can of roast veal apparently was not consumed, because it turned up in a museum 114 years later. Inquisitive scientists opened it and decided to check the effectiveness of the canning process. They were not quite brave enough to try the veal themselves, but the rats and cats that had the pleasure of partaking of the 114-year-old feast not only survived, but thrived!

This now brings us to the problem of the Tin Man in L. Frank Baum's classic, *The Wizard of Oz*. A mystery even more confounding than the Napoleonic pants. When Dorothy firsts encounters him, he is a little stiff, to say the least. He has to be appropriately oiled before he can begin his quest for a heart. This implies corrosion, but tin does not corrode. And it is unlikely that he would have experienced low enough temperatures in Oz to undergo any allotropic conversion. In any case, then, he would certainly have lost some of his shine. So I'm afraid we are left with the conclusion that the Tin Man was really Tin Plate Man. That would explain it all. Scratches in the tin plate (an obvious possibility for an ax-wielding woodsman) would have exposed the iron underneath, causing exposure to air and moisture. This sets up a situation known as "cathodic protection," whereby the thin layer of tin oxide on the surface of the metal is converted to metallic tin as the iron turns to rust! The inside corrodes, while the outside stays shiny!

I thought it worthwhile to check the scientific literature to see if anyone before had addressed the problem of the creaky Tin Man. Apparently not. But I did come across one item that was disturbing, and that may undermine my research with tin buttons. One spoilsport historian claims that only Napoleon's officers had tin buttons and that common soldiers had buttons made of bone, which would have easily stood up to the bone-chilling Russian winter. And what did my own kitchen research show? The tin buttons were unaffected by the cold. So I guess the only thing that crumbles is the story about Napoleon's soldiers and their falling pants.

FIREBOMBS, BEDPANS, AND A MOLDY CANTALOUPE

"Sold, for £23,000!" cried the auctioneer at the famed Sotheby's auction house in London. The elated winning bidder, a representative of the Pfizer pharmaceutical company, rushed forward to claim his prize. It didn't look like much: just a simple glass slide with a few black smudge marks. What made it so valuable was the inscription on the back. "The mold that makes penicillin," it said. And there was also a signature: "Alexander Fleming." Fleming had given this little historic relic to Dan Stratful, his laboratory assistant, sometime after that momentous day in 1928 when he "discovered" penicillin. Pfizer was one of the companies involved in eventually bringing the drug to market and was thrilled to acquire this landmark sample.

Alexander Fleming's accidental discovery of penicillin is one of the most often related scientific anecdotes. Unfortunately, it is usually oversimplified to the point of inaccuracy. Penicillin was certainly not an overnight success. The fifteen years between Fleming's original observation and the commercial production of the drug featured a number of events that would prove to be critical in leading to the world's first "miracle drug." In addition to the research undertaken by a group of scientists at

Oxford University in England, and another at the US Agricultural Research Laboratory in Peoria, Illinois, important roles would also be played by a miner's eye, firebombs, a rosebush, some bedpans, and a moldy cantaloupe!

Fleming trained as a surgeon at St. Mary's Hospital in London but never pursued the profession as a career. That's because he was a crack shot with a rifle! After graduation Fleming, who was looking around for a surgical position, was approached by the captain of the rifle club at St. Mary's, desperate to improve his team. He convinced Fleming to stay and take a position in the hospital's Inoculation Service. It was here that Fleming made his first important discovery. Having personally seen the misery caused by infected wounds in World War I, Fleming began to look for substances that were effective against disease-causing bacteria. One day a teardrop fell into one of his cultures, killing some of the bacteria. Fleming isolated the active ingredient, lysozyme, and realized that since it was found in tears, it was unlikely to harm human cells. It didn't, but it was not effective against disease-causing bacteria, either. The experiment did prime Fleming for his famous "discovery" on September 3, 1928, though. Having just returned from a vacation, he noted a mold growing in a culture dish of Staphylococci bacteria. More importantly, the bacteria around the mold were dead! This mold spore, which had probably drifted in from the mycology lab on the floor below his, was apparently releasing some chemical that was toxic to bacteria.

Within a year Fleming identified the mold as *Penicillium notatum,* coined the term "penicillin" for the active ingredient in what he had first called his "mold juice," and published his account in a highly respected British medical journal, *The Lancet*. He then set his assistants, Frederick Ridley and Stuart Craddock, to the task of isolating the active ingredient from the *Penicillium* mold. They found penicillin to be very unstable and only managed to make crude extracts. At this point Fleming lost some of his interest, especially after noting that penicillin was powerless against the bacteria that caused cholera and bubonic plague. He never did become involved in any human penicillin research. But he did play a role in a penicillin cure! Craddock had gone on to become a country doctor and was often visited by Fleming. He just happened to be there when one of Craddock's patients mentioned that his dog was dying of a foot infection. Fleming sent to London for one of his crude penicillin extracts and applied the powder to the dog's foot. The infection disappeared!

But Fleming's influence did not stop with country canines. Apparently he was a terrible lecturer, so his students had to look up the original papers to which he referred instead of relying on lecture notes. And so it happened that Cecil Paine read Fleming's original paper about penicillin and became totally enthralled. Paine, though, was set on becoming a practicing physician, not a researcher. A couple of years later, while working at the Royal Infirmary in Sheffield, he was asked to see a miner who had lacerated an eye and developed a terrible *Pneumococcus* infection. In those days this usually meant removing the eye, but Paine decided to give Fleming's mold a try. He irrigated the man's eye with a crude extract of penicillin and managed to save his sight. Encouraged by this success, Paine used his preparation to treat the eye of a baby who had contracted gonorrhea from his mother at birth. Again, penicillin did the job!

So why is Paine hardly ever mentioned in accounts of the history of penicillin? Simply because his work was never published in the scientific literature. Later, Paine explained that since he was using a crude extract and had not carried out sufficient experiments, he did not think his work met the criteria for scientific publication. Luckily, though, while at Sheffield University Hospital in 1932, Paine met a newly appointed professor of pathology. In a conversation with Howard Florey, he mentioned his experience with penicillin. The professor seemed to take little interest in this at the time, but the seed, which six years later would sprout into the program at Oxford University that was to alter medical history, had been planted.

In the spring of 1935 Florey was appointed chair of pathology at Oxford University, and had some changes in mind. He wanted to give pathology a "good twist away from diagnosis and morbid anatomy," and place more emphasis on research. Florey himself had a strong research background and while training at Cambridge University had developed an interest in lysozyme, an enzyme in mucus that had antimicrobial properties. This was the research he wished to pursue at Oxford, but the department had no biochemists on staff familiar with the necessary techniques. Florey asked colleagues for recommendations and came up with the name of Ernst Chain, a German who had fled the Nazis and completed a PhD at Cambridge. Hiring Chain would prove to be an inspired choice.

As Florey and Chain were exploring the nature of lysozyme, the bombs began to fall on Europe. Military and civilian casualties piled up, and deaths from infected wounds mounted. Although lysozyme had some interesting antibiotic properties, it was clear that mucus would not be the answer to the infection problem. Now, with the war taking its terrible toll, Florey recalled his meeting with Cecil Paine, and determined to do something Fleming had not managed—to isolate penicillin from the mold. This was no simple task, since the mold produced penicillin only in tiny amounts, and the compound was very labile.

The job of scaling up production to get enough penicillin for testing went to Norman Heatley, a gifted chemist. Heatley found that amyl acetate was an ideal solvent for extracting penicillin from the mold, and more significantly, he found ways to produce the mold in large amounts. Because of wartime shortages in glassware, the inventive Heatley resorted to using food tins, milk jugs, and even bedpans. These turned out to be ideal because the *Penicillium* mold grows on the surface of its nutrient broth, and bedpans, which are easily stackable, provide a large surface area for growth.

Even with production scaled up effectively, isolation of pure penicillin proved to be a challenge. The final product always seemed to be contaminated by other substances that were toxic to test animals. Florey then brought in Edward Abraham, an organic chemist, who not only found a way to remove impurities using the newly discovered technique of column chromatography, but also successfully determined the exact molecular structure of penicillin. The stage was now set for the first critical experiment with pure penicillin.

On May 15, 1940, Florey and Chain infected eight mice with *Streptococcus,* and an hour later, injected four of them with penicillin. The treated mice lived, the others died! "It looks like a miracle" was the exuberant comment from the normally taciturn Florey. But would it work on humans? In February of 1941, the team had a chance to find out.

Albert Alexander, a policeman, had accidentally scratched his face while pruning roses. The infection spread rapidly through his body, and caused the loss of an eye. Florey suggested trying penicillin, and the results were spectacular. The patient showed instant improvement but soon relapsed, as reserves of the drug had been exhausted. Florey's crew even resorted to using penicillin extracted from the policeman's urine, and saw temporary improvement. Unfortunately, the penicillin ran out, and the policeman died. But within a short time, the power of penicillin to save lives was demonstrated in a handful of other patients, and it became clear that industrial-scale production techniques had to be devised. England, though, with a possible German invasion in the offing, was not the place to do it.

Florey and Heatley, with support from the Rockefeller Foundation, traveled to the United States to drum up American support for the project. Their coats were smeared with dried spores of the *Penicillium* fungus in case Britain fell to the Germans while they were away. Florey had contacts at the Department of Agriculture, and by the time he returned to England a few months later, an active penicillin program was under way at the Department's research lab in Peoria, Illinois, with help from Heatley, who had been left behind. Peoria was surrounded by cornfields, prompting the scientists to try corn-steep liquor as the nutrient to grow the *Penicillium* mold. Yields increased dramatically. Then they found that the mold could be grown in submerged cultures in large tanks as long as these were constantly aerated and agitated. And then came the moldy cantaloupe!

The Peoria researchers wondered if some other molds might be more efficient at producing penicillin than Fleming's original one. They searched cheese factories and fruit stands for candidates that would then be tested by Mary Hunt, who quickly became known as "Moldy Mary." Well, one day Moldy Mary found a moldy cantaloupe in a Peoria fruit store. She brought it back to the lab, and guess what! That mold, *Penicillium chrysogenum,* was the best little penicillin factory anyone had ever seen! Mutating this mold with x-rays led to even better strains, and with the help of a number of pharmaceutical companies, enough penicillin was soon produced to treat all the wounded. By D-Day, penicillin was available to the public. As a final footnote, while Florey was still struggling to isolate penicillin, Fleming contacted him on behalf of a friend suffering from meningitis. Florey provided the drug, which Fleming then injected into the man's spine, curing him. Fleming's mold had saved a life! And it would save millions more. Fleming, Florey, and Chain received the 1945 Nobel Prize in medicine.

Radar and Hot Coffee

The scientists working at the Raytheon Company during World War II were undoubtedly highly stressed. They were working on improving radar (Radio Detection and Ranging), the electronic detection system that had been invented by the Scottish physicist Robert Watson-Watt. Lots of coffee must have been consumed during those pressure-packed days at Raytheon, and cups must sometimes have become cold when left sitting on a bench as the research heated up. But not all the coffee cooled down! Coffee in cups that had been left near the electronic tubes used to produce the microwaves needed for the operation of radar actually heated up. The concept of a microwave oven was born!

Microwaves are a form of low-energy electromagnetic radiation. Unlike x-rays or ultraviolet light, they don't have enough energy to break chemical bonds. They do, however, have the ability to interact with molecules that have positive and negative regions. Water is an excellent example of such a "polar" substance, since the oxygen in H_2O has a partial negative charge, while the hydrogen atoms have a positive character. Just like waves in water, microwaves have crests and troughs. The charged regions of a water molecule align differently with the crests and troughs, which means that as a wave passes through a molecule, it begins to spin. These spinning molecules are jammed together so closely that friction is created, and this friction generates heat. Microwave ovens, therefore, work by heating up water. Since most foods contain significant amounts of water, they can be cooked by microwave.

By the 1980s cheap microwave ovens had been developed and began to alter the kitchen landscape. Heating leftovers became a snap, and fresh popcorn was only a couple of minutes away. But it didn't take long for the critics to target the novel technology. Microwaves cause cancer, they said. "You might as well have a nuclear reactor in your kitchen," one health food advocate declared. That is just plain silly! Microwaves do not break chemical bonds, and therefore cannot disrupt DNA—a process essential to carcinogenesis. Furthermore, microwaves are confined to the oven, and there are no "residual" waves attacking unsuspecting consumers as the door is opened. Then there was the allegation that microwaves change the chemistry of the food, making it unhealthy. One anti-microwave advocate declared that the force that causes water molecules to spin also rips apart and deforms the molecular structure of the food so "that it is no longer food; it just looks as though it is." Utter nonsense.

Yes, when microwaved, the chemical makeup of food does change. Any form of cooking has the same effect. Heat initiates a number of chemical reactions, most of which are desirable. Proteins become more digestible, and various flavored compounds are produced. True, microwaved cooking may lead to a less tasty meal, since compounds such as thiazole, furan, and pyrazine—all very flavorful—are not as extensively produced by the lower temperatures and shorter cooking times in a microwave. On the other hand, heterocyclic aromatic amines (HAAs), which have been linked to cancer, are far more likely to form in fried, grilled, or broiled meats than in those that have been microwaved.

Spanish researchers recently caused quite a stir when they compared nutritional losses in steamed, boiled, and microwaved broccoli. They were interested in levels of flavonoids, compounds that have decided anti-cancer effects. Microwaving broccoli resulted in a 97 percent loss of flavonoids, while there were minimal losses associated with steaming. Reports in the lay press inferred that this research demonstrated that microwave cooking destroyed nutrients. It did no such thing. What it did show was that the researchers have no idea about how broccoli should be cooked in a microwave oven. They immersed the florets in water and cooked on "high" for five minutes! Perhaps they like the texture of disintegrated broccoli. I don't. The way to cook broccoli is to put just a couple of spoonfuls of water in the bottom of a glass bowl, add the florets, cover, and cook for two minutes. Had they done this, I'd bet that losses would be comparable to steaming. But that doesn't make for as good a story as "microwaving destroys nutrients," does it?

Is there any real risk with microwave ovens? You bet! They have a long history of causing terrible eye injuries. The problem, considering the extensive use of such ovens, is not huge, but there are at least thirteen cases in the medical literature of people being viciously attacked by microwaved eggs. One of the first such reports appeared in the *New England Journal of Medicine* back in 1991. A nineteen-year-old man, for reasons known only to him, heated seven eggs in their shells in a microwave oven at full power for five minutes. He managed to remove them from the oven uneventfully, but when he sat down at a table, six of the eggs spontaneously exploded, causing severe burning about the face. The problem was pressure buildup due to steam inside the egg. Even eggs cracked into a bowl can be dangerous. Several people have been injured when they pierced the yolk of a microwaved egg with a fork. The cooked membrane around the yolk can sustain a great deal of pressure—at least until it is pierced. Then it retaliates, releasing a jet of steam.

Be careful when microwaving coffee as well. Sometimes the liquid can become superheated without boiling. When the cup is picked up, the coffee can virtually explode out of the cup. So wait until you see the liquid clearly boiling in the oven before handling the cup. Those researchers at Raytheon may have noted this effect, but probably saw no need to comment on it. After all, they would never have guessed how their work on radar would eventually make waves in virtually every kitchen.

Spontaneous Human Combustion

Whenever I hear about solar flare activity, I anxiously wait for reports of people who have spontaneously burst into flames. That's because solar flares have been postulated as a potential explanation for spontaneous human combustion, one of the most bizarre phenomena ever described in the annals of science. Many North Americans first learned about this weird business years ago when the popular TV program *Picket Fences* captured viewers' imagination with an episode in which the town's mayor was found reduced to a pile of smoldering ashes in a room that otherwise shows no signs of a fire. After some discussion and reference to other similar cases, the authorities offered an explanation of "spontaneous human combustion." The implication was that they had witnessed the aftermath of a rare but well-documented event, in which a human being suddenly bursts into flame and is almost totally consumed. Can this really happen?

The controversy over supposed spontaneous human combustion goes back to the nineteenth century. Charles Dickens certainly believed in the possibility. In his *Bleak House*, spontaneous flames consume the sinister, drunken Mr. Krook. This is not simply poetic license. Dickens explains in the preface to the novel that he "took pains to investigate" the subject. The publication of *Bleak House* brought an immediate rebuttal from the famed German chemist Justus von Liebig, who was one of the early champions of the scientific method. Conclusions, insisted Liebig, must be based upon rigorous observation. He took a swipe at Dickens by writing that "the opinion that a man can burn of himself is not founded on a knowledge of the circumstances of death but on complete ignorance of all causes or conditions which preceded the accident and caused it." So who's right, Dickens or Liebig?

Let's investigate the two best "documented" cases of spontaneous human combustion. The story of the mysterious 1951 death of sixty-seven-year-old Florida native Mary Reeser is recounted in virtually every book that deals with "unsolved mysteries." The firemen called to her apartment were confronted by a gruesome sight. All that remained of the portly Mrs. Reeser was a shrunken skull, a pile of ashes, and some charred bones with an intact left foot sticking out. Much of the rest of the room was untouched by fire. A clear case of spontaneous human combustion! Right? Well, not exactly.

An examination of all of the facts in the case, as Liebig would surely have done, leaves us with a decidedly different impression. The victim had been wearing a flammable nightdress and was sitting in an overstuffed armchair, which had also been reduced to ashes. She was a smoker and regularly took sleeping pills. So what happened? In all likelihood, she fell asleep, dropped her cigarette, and was consumed by the flames fueled by the chair stuffing and her own body fat. The reason the fire had not spread was also evident to investigators. The floor of the apartment was concrete! Furthermore, unlike the picture painted in the mystery books, there actually was heat damage to the ceiling, and a nearby table and lamp were destroyed. The ceiling, draperies, and walls were coated with a smelly, oily soot characteristic of burnt fat.

A very similar scenario confronted firefighters summoned to the apartment of Dr. John Irving Bentley in a small Pennsylvania town on December 5, 1966. The semi-invalid, ninety-two-year-old retired physician lived alone, and used a walker to get around. His final trip appears to have been to the bathroom, because that's where the walker was found, tilted over a gaping hole where the floorboards had burned through. Next to the hole were a pile of ashes and the remains of the right leg, intact from the knee down, brown but not charred. Bentley was a pipe smoker and possibly ignited his dressing gown while trying to light his pipe. He then hobbled to the bathroom to try to put out the flames, but obviously didn't manage to do so.

In both of these cases there appears to have been a reasonable source of ignition, so the combustion was not "spontaneous." But it was still mysterious. How could the human body be consumed by flames so totally and not set everything around it on fire? The earliest theory, propounded in the eighteenth century, suggested that the cause of such conflagrations was alcohol. The victims were imbibers who saturated their body with alcohol, which somehow, perhaps as a result of divine retribution, burst into flames, consuming them from the inside. Experiments were designed to study this theory. Rats, soaked in alcohol for up to a year, were exposed to the air. The pickled rodents did not spontaneously ignite and even when they were set on fire did not mimic "human spontaneous combustion." While the skin flared up impressively, and there was charring of the outer flesh, the internal organs and bones were unaffected. So if alcohol is not the answer, what is?

When scientists can't offer an explanation to a mystery, proponents of outlandish theories rush in to fill the void. They present selected facts and then offer to explain the supposed scientific paradox with ideas ranging from the sublime to the ridiculous. Care to hear a few? How about "demonic possession that is thwarted by incompatible psychic energies in the victim, causing a violent explosive reaction?" Or "an internal subatomic chain reaction" caused by "subatomic particles called pyrotrons"? That is the pet theory of Larry Arnold, a former Pennsylvania school bus driver who has become an "expert" on spontaneous human combustion and has written extensively on the subject. Needless to say, science knows of no "pyrotrons." Then there is the most intriguing theory that HSC is caused by "alien forces beaming otherworldly fire at humans for experimental reasons"!

For some spontaneous combustion devotees, those alien forces are not generated by little green men, but by our large yellow sun. Sun flares, they claim, flood the earth with protons and electrons, which cause changes in the earth's magnetic field, which in turn can destabilize molecules in our body and cause electric discharges that momentarily turn the stomach into a microwave oven. The unfortunate victim then cooks himself from the inside out. Weird stuff, to say the least.

There is really no need to consider such remarkable theories when smoldering cigarettes, nearby fireplaces, and combustible materials can explain the apparently bizarre conflagrations. Of course we would like to have proof of this. And a British TV documentary seems to have furnished it. After reviewing the most famous cases of spontaneous combustion, scientists on the show proposed that cremation temperatures are not necessarily required to consume a body. If body fat liquefies and burns slowly, a "candling" effect is produced whereby the combustion products are drawn straight up instead of spreading sideways. This would explain the oily deposits on the ceiling. They then proposed to prove their theory.

A deceased pig was dressed in human clothes and placed in a household living room complete with carpets, curtains, and TV. Its clothes were then set on fire with a match. A slow burn followed, and the candle effect was clearly noted. After a few hours the body, including the bones, was completely consumed—

except for a few pig knuckles. Just like in the reported cases of spontaneous human combustion, this "spontaneous" pig combustion left the surfaces in the room covered by fine, oily soot. Some plastic articles melted, but the furnishings in the room were unaffected.

It remains important to investigate each case of reported SHC individually. But why not consider the fascinating scientifically established principles of combustion before invoking demons, nonexistent subatomic particles, or solar flares as being responsible for the strange effect? So far, I haven't heard of anyone bursting into flames after reports of solar activity. But some of the silly stuff written about the consequences of such solar flares does cause me to do a slow burn.

Forceful Sole Searching

I got my first pair of Florsheim shoes when I was thirteen years old. It seemed an appropriate bar mitzvah present because, after all, I was becoming a man. I had earned the right to wear an adult shoe. It squeaked a little, as I recall, but it was a very good shoe. But a few years ago, the Florsheim company did a little squeaking itself. The company was sued by a consumers' rights group in California called the Consumer Justice Center for false advertising and consumer fraud. What's going on here? How can something as simple as a shoe become embroiled in a courtroom drama?

Well, it can if the shoe is a golf shoe and claims to do more than just provide a comfortable barrier between the foot and the ground. Questions naturally arise when claims are made about increased circulation, reduced foot, leg, and back fatigue, pain relief, and improved energy levels. That seems to be quite an accomplishment for a shoe. Ah, but it's no ordinary shoe.

Florsheim's "MagneForce" has magnets built right into it. And therein supposedly lies the magic. But according to the Consumer Justice Center, there is no magic here, just some trickery.

Magnets are very popular as healing tools these days. There are magnetic mattresses, pads, bandages, insoles, rings, and bracelets. You can even buy magnetized water. A remarkable Web site sells "immortality rings" that claim to increase life span. The inventor, American Alex Chiu, offers up incomprehensible equations and diagrams to buttress his claims of having solved the problem of aging.

Perhaps I am just not smart enough to understand Chiu's explanations and diagrams because I'm not wearing the immortality rings. You see, they also boost your IQ to 180! I guess Chiu must wear them all the time, because now that he has solved the problem of immortality, he has gone on to other things. He has invented a teleportation machine. He ensures us that he is "not one of those stupid morons who doesn't know what he is doing." Why teleportation? Because when we are immortal we will have plenty of leisure time, which we will be able to use to pop up here or there.

Admittedly, magnets can produce fascinating effects. The idea of an invisible force that attracts iron is mind-boggling. And without magnets we would have no electric motors, tape recorders, VCRs, or indeed credit cards to pay for them. But using magnets for healing is another matter altogether. Unfortunately, very scientific-sounding claims about healing abilities can be made, and believed, by people who do not have a good grasp of magnetism. This, of course, means most people. There is a pattern to these claims.

Usually it all begins with a reference to some form of ancient wisdom. Like how Hippocrates, that most famous of all ancient doctors, used magnets to heal the sick. Or how Cleopatra wore magnetic jewelry to preserve her youth. The fact is that neither the ancient Greeks nor the Egyptians ever used magnets in this way. But what if they had? They did many senseless things. Hippocrates, for one, believed that a mixture of horseradish and pigeon droppings could be used to treat baldness. Anyway, after supposedly having established the long and fruitful history of magnetic therapy, the scene often shifts to those flag-bearers of our future: those modern knights, the astronauts. The story is that magnets incorporated into spacesuits resolved many of the astronauts' health problems. And a story it is. No magnets have ever been incorporated into spacesuits for this purpose.

But the real "scientific" selling point revolves around the so-called "electromagnetic nature of the human body." There is usually talk of how our nervous system relies on small electric currents and how MRI machines diagnose disease by examining changes in magnetic fields inside the body. Both of these are true. But then from these observations we are asked to conclude that applying small magnets to the body can treat ailments. A scientific and logical non sequitur. First of all, the electricity being talked about really involves the flow of small charged particles called ions. Their motion could in theory be affected by giant magnetic fields, but not by the small magnets sold for healing purposes that have strength in the range of refrigerator magnets. Even magnets used in Magnetic Resonance Imaging (MRI), which are hundreds of times stronger than the healing magnets, do not affect the nervous system, and have no effect on blood flow. It isn't surprising that there is no effect on blood flow. While magnet advocates maintain that blood flow is affected because hemoglobin contains iron, the fact is that the iron it contains is not magnetic. And that's lucky, isn't it? We wouldn't want our blood to be ripped out of our body when we're undergoing an MRI scan.

And if magnetic fields can heal, shouldn't we have reports of people being healed, or at least being energized, after an MRI scan? The huge magnetic field this instrument generates certainly penetrates the body. Not like the little healing magnets. Those in a shoe produce fields that may penetrate the sock but not much else.

There is another common claim used to buttress sales. The claim is that we suffer from magnetic deficiencies. Physicists say, at least according to the magnet salespeople, that the earth has lost some of its magnetism, and since human evolution occurred in higher fields, we are now feeling the ill effects of the reduced magnetism. First of all, physicists do not say this, and even if they did, it would not mean that there is a related health effect. As it is, the earth's magnetic field varies tremendously. At the poles, it is 0.6 gauss, double that at the equator. No one has ever noted any variation in disease patterns based on magnetic field geography. Other related claims are barely worth refuting. Including the one about the earth's magnetic core. It goes like this: "The earth itself is a giant magnet with north and south poles and a liquid core. (True enough). The hot liquid creates a magnetic field, which at the earth's surface is relatively weak (still true), but serves to keep humans attached to the earth's surface. Without this magnetic field, we would spin into outer space." This is absurd. As any grade one student knows, gravity, not magnetism, keeps our feet firmly planted on the ground.

It would seem, then, that the arguments used to promote the sales of magnetic healing products are not scientifically justifiable. But that does not rule out their possible effectiveness. Perhaps magnets do perform the wonders their advocates claim to have experienced, but through some completely different mechanism. That's why the only way to study efficacy is through controlled trials. And what do these show? Not much. Although dozens of studies have been carried out, there is only one that is constantly quoted as having shown a positive effect.

And that was in a rather rare condition known as post-polio myalgia, and has not been reproduced. But trials of magnetic jewelry, and even insoles, have shown no benefit. And that is exactly why Consumer Justice Center sued Florsheim shoes. For making claims that are not scientifically supportable. I wish I could tell you otherwise. I wish that all those people who tell me about their wonderful experiences with magnets were reporting something more than just a placebo effect. Believe me, if any compelling evidence emerges, I'll be happy to relay it.

In response to the lawsuit, Florsheim has greatly pared back its claims about the $200 MagneForce on its Web site. Maybe Florsheim should drop the dubious health claims, forget the magnets, and get Tiger Woods to wear the shoe. People would buy it then even if the spikes were on the inside. That's the kind of magnetic pull Tiger has.

MAGICAL HYDRIDES

The flyer placed under my windshield wiper in the shopping center parking lot informed me that I had been specially selected (curiously, along with every other shopper) to receive an audiotape about an astounding scientific breakthrough. A "leading" scientist, a "Nobel Prize Candidate" (a nonexistent distinction), had discovered the secret to longevity, a secret that would be revealed on the tape that would be sent to me if I called the number provided. Why not? There are worse things than living to 120. That's the age, according to the flyer, within our grasp. And so began my journey into the mysteries of "Microcluster Technology"—the "Nutritional Breakthrough of the Century"—and its inventor, Dr. Patrick Flanagan.

We first encounter seventeen-year-old Flanagan in the pages of *Life* magazine in 1962, hailed as "one of the hundred most important young men and women in the United States." At the age of eight, he had a dream in which he was told to learn all about physics and electronics in order to help people. So he began to tinker at home and invented the "neurophone," a device that he claimed would allow deaf people to hear, and also suggested that the same technology would one day allow blind people to see. The brash claims, quirky personality (Flanagan routinely stood on his head to help him think), and cocky nature caught the attention of the *Life* reporter, who said that the clever youngster could soon be looking at a million dollars from his inventions. The reporter was half right. The deaf are still deaf, and the blind are still blind, but Flanagan is a millionaire.

It seems, however, that it was not extensive formal education that paid off. Indeed, although he is widely referred to as "Dr. Flanagan MD," it turns out that the MD refers to some ambiguous credit abbreviated as "Med. Alt." He is no medical doctor. Insight, it seems, into the secrets of life came not at a university, but rather inside the Great Pyramid. It was while he wandered through the chambers there that he "discovered" the key to health. The magical dimensions of the pyramid somehow infused the body with "Zero Point Energy" that had remarkable preventative and curing properties. His 1973 book, *Pyramid Power,* was an amazing piece of work. It described how animals, plants, and humans could all benefit from the magical properties of pyramids. I recall the Toronto Maple Leafs' attempt to avail themselves of this phenomenon by placing little pyramids under the players' bench. It didn't do them much good. They may have had Pyramid Power, but the Canadiens had "Flower Power." Guy Lafleur was too strong a force for the pyramids.

Flanagan sold more than books. He sold "revolutionary flat pyramid power pendants." He asked us to imagine the dimensions of the Great Pyramid "screwed down flat from the top until it looks like a crop circle." We were further comforted to learn that "all the energy collecting and focusing ability of the pyramid remains with the additional benefit of being able to wear it on your body all day. The only Zero Point Energy device that is safe for continuous wear." No more worries about those other Zero Point Devices produced by competitors, which obviously put our health at risk.

The Zero Point Energy device was not Flanagan's greatest invention. That label has to be reserved for "Microhydrin Nanocolloidal Mineral Hydride," which, according to the tape I received, is the secret behind health and longevity. But the amazing stuff has more mundane uses as well. If you slam a door on your finger, all you have to do is take some Microhydrin and the blood blister will disappear. This comes from no less an authority than Dr. Flanagan, who apparently makes a habit of slamming doors on his fingers. Oh, by the way, he also managed to cure his dog of a degenerated hip with his miracle supplement.

The story behind Microhydrin is even more captivating than Pyramid Power. It seems that way back in the 1960s, a Romanian physicist, Henri Coanda, discovered that snowflakes were alive. As Flanagan recounts, Coanda found that "snowflakes have a circulatory system like a living being; they have little tiny veins like little arms, frozen on the outside; water circulates in the center of them, and the life of a snowflake is as long as the water flows in the veins or until it freezes, and then it dies." Coanda measured the life of snowflakes in different areas of the world (that must have been some grant application he wrote), and discovered that he could predict within five years the average death of people in a given area by just examining the water. There were five areas where people routinely lived to well over 100 (birth certificates please!), and they all drank special water that came from glaciers. Coanda could not solve the mystery of this water but gave his research to the child prodigy, Flanagan, whom he was convinced would one day create a machine that would make "Hunzaland water." Hunza is a region in the Himalayas with legendary longevity. It is also notorious for its poor record keeping.

So Flanagan began his flaky research, which after some thirty years culminated in the discovery that glacier water in Hunza had tiny clusters of minerals that have a very high "zeta potential." It is unclear how this ties in to the living snowflakes, since these are composed essentially of frozen rain, which has no mineral content. In any case, the mineral clusters in Hunza water change the way water molecules are attracted to each other and create a "quasi liquid crystal structure." According to Flanagan, "when we drink ordinary water, which has no structure, our bodies have to convert that water into living water which has crystalline structure. . . . Hunza water already has that structure, so it is already biological." What makes it biological? It has a lower surface tension. It spreads more easily. This is what the body needs. How do we know? The proof, Flanagan says, is in mashed potatoes. You can't mash potatoes with room temperature water because the water doesn't wet the potatoes, and they form clumps. Heated water has a lower surface tension and is wetter and can be used to mash potatoes. This we are somehow supposed to relate to health. I kid you not. I haven't been blessed with enough imagination to make this stuff up.

Since most of us can't get fresh Hunza water delivered from the Himalayas, Flanagan has created a "complete analogue." All we have to do is swallow some capsules (and some truly convoluted science) which will reduce the surface tension of our internal fluids and which will also provide us with the ultimate magic in Hunza water, namely "negative hydrogen ions." We are all deficient in negative hydrogen, as Flanagan says, and therefore prone to disease. Negative hydrogen, or "hydride," you see, is the ideal antioxidant. It neutralizes free radicals. It can also reverse cancer. This time Flanagan quotes a French authority who claims that "cancer cells have no hydrogen in them. . . . [T]hey are the only cells in the body that have no hydrogen in them." Of course, the microcolloids in Mycrohydrin, based on food grade silica ("sand" in common language), can restore our negative hydrogen balance. I guess I better hit my chemistry books again because I'm totally baffled by Flanagan's discussions of bio-terrains, reduction potentials, and hydrogenated microclusters—in spite of the fact that I've had a fair bit of experience working with hydrides in the laboratory.

Maybe my confusion stems from thinking too much about how these magical hydrides would withstand the onslaught of hydrochloric acid in the stomach. Flanagan says that when people think too much they produce free radicals, and "often people can't get rid of those free radicals, so their brains are actually becoming toxic and damaged." This can even cause depression. The secret to mental clarity therefore lies in Microhydrin or, perhaps, thinking less. I guess Flanagan must also think too much. How else could he come up with the plethora of pseudo-scientific garble that oozes from his mouth? Maybe it comes with such ease because the microcolloids have reduced the surface tension of his fluids. Of course, if surface tension were the secret of life, a little soapy water would do the trick. This would be a lot cheaper than "nanocolloids." Maybe "Dr. Flanagan" could investigate its beneficial properties. He could start by using it to wash his mouth.

Natural Cures "They" Don't Want You to Know About

Funny how the mind can sometimes make strange associations. I was enjoying the incredible recuperative powers of Jack

Jeebes in *Men in Black II* when images of Kevin Trudeau popped into my mind. Jeebes, you should know, is the slimy alien pawnshop owner whose head immediately regenerates after it is blown off. And who is Kevin Trudeau? Well, he is the king of infomercials, a man who has shamelessly hawked dozens of questionable products on TV programs cleverly disguised to look like documentaries. No matter how many times he is fined or reprimanded, the man who spent two years in federal prison for credit card fraud just keeps popping up again and again to dupe the public and annoy scientists. Sometimes he plays the skeptical reporter interviewing some expert about the latest miracle breakthrough; sometimes he himself is the "expert." One of his classic infomercials featured the Mega Memory course designed by "Kevin Trudeau, memory expert" and founder of the "America Memory Institute." Well, I have my suspicions about the effectiveness of this course, seeing that the "memory expert" seems to have trouble remembering legal proceedings that have been brought against him.

In 1998, the Federal Trade Commission (FTC) in the US, which looks after consumers' welfare, charged that Trudeau made false and unsubstantiated claims on behalf of various products on his infomercials. The charges were settled when Trudeau agreed to pay a fine of half a million dollars and promised to abstain from making false claims. But it seems the lure of huge profits was too much, because in 2003 the FTC again brought proceedings against Trudeau, after one of his programs featuring "expert guest Robert Barefoot" claimed that "Coral Calcium" was a cure for a wide array of human ailments. In a separate infomercial Trudeau falsely insinuated that a strip of "Biotape" applied to the body could provide permanent relief from severe pain. This time, to settle the charges, the prolific marketer agreed to a fine of $2 million and a permanent ban from appearing in, producing, or disseminating future infomercials that advertise a product, service, or program that cannot be backed up scientifically.

Since the FTC certainly has no desire to curtail the freedom of speech, infomercials for books or newsletters were exempted from the judgment. And that was just the opening Kevin Trudeau needed to launch another profit-making venture, one that is even more disturbing than his other schemes. This time he is pushing his book *Natural Cures "They" Don't Want You to Know About*. The "They" of course refers to the FTC, the Food and Drug Administration, associations like the American Cancer Society, and pharmaceutical companies, all of whom, according to Trudeau, want to undermine the health of the public for their own greedy motives. In fact, it is this nonsensical work that may undermine health. It certainly sent my blood pressure soaring!

Trudeau is not one for understatement. He boldly declares at the outset his contention that there are all-natural cures for virtually every disease and ailment. Of course, these cures are suppressed by the medical establishment, lest they cut into profits. To back up his claim, Trudeau regales us with the saga of the maverick (unnamed) scientist who found a herbal cure for diabetes but was paid $30 million by a pharmaceutical company not to market it. Oh, yeah? Where? When? We're also told that the American Cancer Society has swept information under the carpet about a plant that cures cancer in one week. Really? What is that plant? Well, you won't find that information in Trudeau's book, since he claims he is not allowed to talk about specific products because the FTC may then prosecute him and burn his books. Nonsense! But this clever man suggests that he has found a way around government harassment by directing people who want the specific information to his Web site. Don't expect to find out about the miraculous cancer-healing plant just like that, though. You'll have to register and pay a fee. Since I was unwilling to contribute to Mr. Trudeau's already copious coffers, I'm afraid I will remain in the dark about this cancer cure.

The drug companies, Trudeau maintains, design medications with side effects so they can then sell you more drugs to treat the new problems that arise. The food industry knowingly puts additives into food that will make people hungry, fat, addicted, and depressed. Pharmaceutical companies, which supply some of these additives, can then sell their antidepressants, which of course have side effects that have to be treated . . . and so on and on it goes. This is just silly stuff. But then there are Trudeau's scientific absurdities. Like that animals in the wild do not get sick. Nonsense. Or that every single person who has cancer has a pH that is too acidic. Nonsense. Pasteurization kills all living natural enzymes. More nonsense. Enzymes are not alive. How about this gem? If you eat an apple today, it has one-fifth the nutrition of an apple of fifty years ago. And even more nonsense. Irradiation changes the energetic frequency of a food, giving the food a frequency that is no longer life sustaining. Thinking the correct thoughts actually changes a person's DNA. Mind-numbing nonsense.

Then comes the advice for good health in a chapter blatantly entitled "How to Never Get Sick Again." Didn't work for me. Made me sick with suggestions like "rebalance your body with frequency generators." ("These machines neutralize the frequency of the disease.") Get a water cooler that adds oxygen to the water. Get fifteen colonics in thirty days. (May not be a bad idea for Mr. Trudeau himself—some elimination seems appropriate.) Use magnetic toe rings. Stay away from electric tumble dryers ("produce positive ions that suppress the immune system"). Have sex often. Well, even the scientifically challenged occasionally get something right.

INDEX

Abraham, Edward 256
Acetone 160
Acetylcholine 132, 163
Acrylamide 25–29
Acupuncture 139–44
Adenosine 65
Adenosine triphosphate (ATP) 95
Adrenaline 121
Aflatoxins 27
Agutter, Dr. Paul 164, 165
Alcohol 61, 181–84, 263
Alcohol dehydrogenase 169, 170
Alcor Life Extension Foundation
    165, 167, 168
Alexander, Albert 256
Alkyl ketene dimers 226
Alkyl succinic anhydrides 226
Alpha-linolenic acid (ALA) 16, 20
Alpha-tocopherol 105–7
Aluminum 117, 210–13, 241
Aluminum chlorhydrate 129
Aluminum hydroxide 128, 219
Aluminum stearate 235
Aluminum sulfate 219
Alzheimer's disease 107–8, 135
Amalgam 195
Amaranth 78–79
Aminolevulinic acid 180

Aminophenol 177
Ammonia 89, 175, 176, 199, 219
Ammonium carbonate 219
Ammonium nitrate 199, 201–4
Amoxicillin 55
Amphetamine 186
Amyl acetate 255
Amyotrophic lateral sclerosis (ALS)
    133
Anabolic steroids 186
Anafranil 154
Anderson, Richard 40
Androstenol 87
Androstenone 87
Anemia 187
Angiotensin-converting enzyme
    (ACE) 50
Anthocyanins 23, 214
Anthrax 123–24
Antibiotics 90
AntiCancer 179
Antifreeze 168, 184, 185
Antioxidant(s) 11–13, 23, 64–66, 86,
    100–102, 107
Antiperspirants 128–31
Apolipoprotein E (Apo E) 74–75
Appert, Nicholas 250
Apples 11–12, 39, 41

Arecoline 132
Arnold, Larry 264
Aromatase 21
Asimov, Issac 14
Asparagine 28
Aspartame 43
Atherosclerosis 23
Atropine 163–65, 190
Autism 75

B vitamins 103
Bacillus thuringiensis (Bt) 85
Bakelite 235
Baking soda 218
Barefoot, Robert 274
Bauxite 211
Bedford, James 166
Behavioral therapy 155
Belladonna 163, 164
Bentley, Dr. John Irving 262
3,4-Benzypyrene 160
Beta-carotene 76–78
Beta-methylamino-L-alanine
    (BMAA) 134, 135
Betel nuts 132
Bile acids 19
BioCure 51
Biotape 274
BioZate 50
Bitter melon 40
Bjelakovic, Goran 100, 101
Black dermographism 215
Black, Samuel 60
Bleaching 224
Blood doping 187
Blood pressure 33, 36, 50, 51
Blueberries 21, 23, 24
Boar Mate 87
Boar-taint 87
Boron 241
Bottled water 157–59
Boyle, Robert 182

Breast cancer 12, 21, 22, 29, 129–31
Broccoli 32, 259, 260
Bt toxin 68
Bubbles 216
Bubonic plague 254
Buchanan, Dr. Robert 164
Byrd, Randolph 119

Cafestol 64
Caffeine 40, 62, 63, 65
Calabrese, Ed 160, 161
Calcitriol 91
Calcium 90–93, 172, 173
Calcium carbonate 91, 226
Calcium fluoride 172, 173
Calcium gluconate 172
Calcium hydroxide 132
Calcium oxalate 169
Campylobacter jejuni 56–57
Cancer 26, 27, 29, 63, 64, 72, 76,
    84, 112, 114, 115, 117, 118, 178,
    273
Capone, Al 182
Carbon dioxide 219
Cartier, Jacques 110
Castor beans 136, 137
Cathodic protection 251
Caudate nucleus 152, 155
Celluloid 235
Cellulose 221, 224
Chain, Ernst 255, 256, 258
Chamorro 134
Charlemagne 15
Cheesecake Factory, The 97–100
Chile saltpeter 70, 197, 199
Chiu, Alex 266
Chlorine 224
Chlorine dioxide 224
Chlorogenic acid 65, 72
Chlorophyll 46, 49
Chocolate 33–35
Cholera 254

Cholesterol 15, 16, 21, 22, 24, 30, 31, 35, 42, 64, 67, 74, 75
Cholinesterase 163
Christie, Agatha 162, 189
Chromium 41
Chymosin 68
Cigarettes 148
Cinnamon 27, 39–42
"Cis" bonds 30
Citric acid 218
Clark, Hulda Regehr 116–18
Cleopatra 266
Clomipramine (Anafranil) 154
Clonidine (Catapres) 154
Clostridium botulinum 70
Clostridium difficile 124, 125
Coanda, Henri 271
Cocoa 33, 34, 36
CocoaVia 34
Cod liver oil 49
Coffee 13, 42, 62–64, 66, 258, 260
Colds 111
Colton, Frank 232
Compulsion 150
Copper 235
Copper sulfide 216
Copper sulfate 207–10, 214, 235, 236
Coprolalia 153
Copropraxia 153
Coral Calcium 274
Corned beef 73
Cornstarch 38
Cortisol 121, 126
Cortisone 230, 231
Cotton 222, 223
Coumaric acids 72
Cowpox 245, 246
Cows 125–26
Craddock, Stuart 253, 254
Cream, Dr. Neill 190, 191
Crohn's disease 51

Crooked House, The 162
Cryolite 212
Cryonics 165–68
Cryoprotectant 167
Cupric chloride 66
Curcumin 12, 13, 76
Cutting boards 57
Cyanide 160
Cyanobacteria 135
Cycad tree 134
Cyclamates 43
Cyclooxygenase 76
Cystic fibrosis 48
Cytokines 111

Darbre, Dr. Philippa 129–31
DDT 161
Dehydroepiandrosterone (DHEA) 158
De la Tourette, Gilles 153
Delta-9-tetrahydrocannabinol 148
Deodorants 130–31
Diammonium phosphate 204
Diatomaceous earth 244
Dickens, Charles 238, 261, 262
Diethyl zinc 225
Diethylhexyladipate 156, 157
Diethylhydroxylamine (DEHA) 155–58
Dimercaprol 195
Dimethyl sulfide 224
Dioscoria 228
1,4-Dioxane 159
Diosgenin 228, 229
Dioxin 161
Dircks, Henry 238
Djerassi, Carl 231–33
Docosahexaenoic acid (DHA) 16–18, 20, 88
Donaldson, Thomas 168
Dopamine 65, 154
Doping 186

Doyle, Sir Arthur Conan 220
Drebbel, Cornelis 206
Dronabinol 148, 149
Durand, Peter 250, 251
Dynamite 242–45

E. coli O157:H7 85, 201
Echinacea 13
Edison, Thomas 233–36
Eggs 15, 16, 18, 20
Eicosapentaenoic acid (EPA) 16–18, 20
Electrolysis 235
Ender, Kornelia 186
Endorphins 143
Enovid 232
Enzymes 45–48
Epilepsy 148
Epinephrine 133
Erythropoietin (EPO) 187, 188
Estradiol 227, 231
Estrogen 21, 22, 232
Estrogen mimics 159
Estrone 230
Etchings 172, 174
Ethanol 27
Ethylene glycol 168–71
Etminan, Dr. Mahyar 178
Ettinger, Robert 166
Eumelanin 176

Fenugreek 40
Ferrous sulfate 192
Fertilizers 84, 197–204
Fire extinguishers 219
Fish oil 17, 20
Flanagan, Dr. Patrick 269–73
Flavonols 34–36
Flavonoids 23, 65, 86, 259
Flax 15–20
Flaxseed 16, 18, 40

Fleming, Alexander 252–54, 257, 258
Florey, Howard 255–58
Florsheim 265, 266, 269
Fluoxetine (Prozac) 154
Fluvoxamine (Luvox) 154
Foam(s) 216–220
Folic acid 103
Fomepizole (4-methylpyrazole) 170
Formaldehyde 71
Fracastoro, Girolamo 123
Free radicals 100, 206
French fries 25
French Harmless Dye Company 175
French paradox 59–62
Frequency generators 276
Fructose 38–39
Fruit-eating bats 133–34
Fuel oil 203
Fumonisin 86
Functional foods 49, 50
Functional magnetic resonance imaging (fMRI) 144
Furan 259
Fusarium 85

Garlic 213–15
Gehrig, Lou 133, 135
Gelatin 222
Gelignite 242
Genetic modifications 20
Genetically modified crops 77
Genetically modified food 67–68
Genetically modified potato 78
Germs 122
Ghrelin 39
Glaucoma 149, 162, 163
Glucose 19, 28, 38–42
Glucose isomerase 38
Glycerine 243

Glycolic acid 169
Godfrey, Arthur 242
Godin, Gérald 114, 115
Gold 208, 213, 215, 216
Golden rice 76–78
Gorilla 236, 237
Goss, Dr. Paul 19
Gourmand syndrome 152
Granholm, Dr. Anne-Charlotte 31, 32
Green onions 58
Guam 133, 134
Guano 197, 198
Guar gum 241
Guarneri, Guiseppe 239
Guillain-Barré Syndrome (GBS) 56, 57
Gum 53, 54
Gutenberg Bible 220, 222
Gutta percha 173
Gymnema 40

Haber, Fritz 199
Hair dye 175–80
Haldol 154
Hall, Charles Martin 212
Haloperidol (Haldol) 154
Heart attack 60–61
Heart disease 21, 64, 74
Heatley, Norman 255–57
Helicobacter pylori 52
Helium 210
Hemoglobin 49
Hemp 222
Hensel, Andreas 52
Hepatitis A 58
Heroult, Paul-Louis-Toussaint 213
Heterocyclic aromatic amines (HAAS) 27, 71, 72, 259
High fructose corn syrup (HFCS) 36–38
Hippocrates 14, 266, 267

Hitler, Adolf 190
Hollenberg, Dr. Norman 33, 34, 36
Holmes, Sherlock 220
Homocysteine 103
Hoodia gordonii 94–96
HOPE 106
HOPE-TOO 105
Hormesis 160, 161
Howitz, Konrad 59, 60
Hunt, Mary 257
Hunter, Dr. John 247
Hunza water 272
Hutchins, Carleen Maley 240
Hydride 273
Hydrocortisone 231
Hydrofluoric acid 172–75
Hydrogen 30, 199
Hydrogen fluoride 172, 174
Hydrogen peroxide 175, 176, 180, 226
Hydrogen sulphide 89

Indian Hemp Drugs Commission 146, 147
Inhoffen, Hans 231
Insulin 38, 40, 41, 162
Insulin receptor kinase 41
In-vitro fertilization 121
Ioannidis, Dr. John 13
Iodine 197
Iron 250, 251
Iron deficiency 192
Isoalliin 214

Jack the Ripper 191
Jenner, Edward 245, 247, 248
Jesty, Benjamin 245, 246, 248

Kahweol 64
Kaldi 62
Kaolin 225
Kaptchuck, Ted 142

Katan, Martijn 30
Keen, Dr. Carl 34
Keller, Friedrich 223
Kent, Dora 167
Kent, Saul 167
Kool Aid 179
Korean ginseng 40
Kuna Indians 33

Laboratorios Hormona 229
Laccifer lacca 52
Lactase 48
Lactic acid 169
Lactobacillus helveticus 51
Lactobacillus GG 124
Latham, James 246
Laughing gas 202
Laval University 91
Lavoisier, Antoine 206
Lead acetate 179, 180
Lead sulfide 180
Legumes 199
Lehmann, Frederick 229, 230
Leibovici, Leonard 119, 122
Leptin 38, 39
Leukemia 82, 230
Li, Hing Hua 127
Lightbulb 174
Lighthouses 192
Lignans 18, 19
Lignin 223, 224, 226
Lind, Dr. James 110, 111
Linen 222, 223
Lobo, Dr. Rogerio 121
Longevinex 61
L'Oréal 176
Luvox 154
Lycopene 86
Lysine 78, 89
Lysozyme 253, 255

Maccaferri, Mario 241
Macular degeneration 108
Magnetic Resonance Imaging 267
Magnets 266–69
Maltodextrin 44
m-Aminophenols 177
Manganese dioxide 204, 205
Manure 196–201
Marijuana 145–49
Marker, Russell 226–33
Markov, Georgi 135,137
McCormick, Katherine 232
McGrath, Dr. Kris 131
McVeigh, Timothy 203
Mega Memory 274
Melzack, Ronald 143
Meningitis 55
Merck 230
Mercuric oxide 205, 206
Mercury 16, 191–95
Mercury fulminate 244
Meridians 141, 144
Meringue 218
Methanol 181–85
Methionine 78, 89
Methyl mercaptan 224
Methyl-t-butyl-ether (MTBE) 184,
    185
Metronidazole 124
Microbes 55, 56
Microhydrin
Microwaves 258–61
Miller, Dr. Edgar 104–6
Montagu, Lady, Mary Wortley 246
Morphine 163, 164
Mousse 218
Mozart 126–28
m-Phenylenediamines 177
Multiple sclerosis 149
Multivitamin 103
Mum 128
Music 125–27

Mycrohydrin 271, 273
Myelin 56
Myoglobin 70
Mysterious Affair at Styles, The 189

Naessens, Gaston 112, 113
Naessens's regimen 114
Nagyvary, Joseph 239–42
Naloxone 143
Nanocolloids 273
Napoleon 248, 249, 250, 252
napthols
Natural Cures "They" Don't Want You to Know About 273, 275
Nauru 198
Negative hydrogen ions 272, 273
Neurophone 270
Nitrates 84, 89, 200
Nitric acid 243
Nitric oxide 36
Nitrogen 196–201
Nitroglycerine 243, 244
Nitro-phenylenediamines 177
Nitrosamines 70–72
Nitrosyl myoglobin 70
Nitrous oxide 200
Nixon, Richard 139
Nobel Prizes 245
Nobel, Alfred 242–45
Nobel, Immanuel 244
Norephinephrine 132
Norethindrone 231–33
Norethynodrel 232
19-Norprogesterone
Norwalk-like virus 58
Nurses' Health Study 16, 31
Nutrigenomics 73, 76

O'Shaugnessy, Dr. Denise 36
Obsession 150
Obsessive-compulsive disorder (OCD) 150–54

Okinawa 98
Oklahoma City 203, 204
Okra 52
Oleic acid 31
Omega-3 50
Omega-3-eggs 17, 18
Omega-3-fats 16–18, 20, 49
Onions 40
Opsin 76
Oral cancer 132
Organic agriculture 83, 84, 198
Organic farms 201
Organic foods 69
Organophosphates 80, 81
Ortho Pharmaceuticals 233
Otitis media 54
Oxalic acid 169
Oxygen 185–87, 204–7, 224

Paclitaxel 13
Paine, Cecil 254, 255
Palytoxin 137
p-Aminophenols 177
Pancreatic enzymes 48
Paper 220–26
Papyrus 221
Parabens 129, 130
Paracelsus 160
Para-phenylene-diamine (PPD) 175–77
Paraquat 80
Parathion 81
Parchment 221, 222
Park-Davis 233
Parker, Terry 148
Parkinson's disease 65, 82, 85
Paroxetine (Paxil) 154
Partially hydrogenated fats 30
Pasteur, Louis 123–24
Pastrami 73
Pauling, Linus 108, 109, 111
Paxil 154

PCBS 16
Pelouze, Theophile 243
Penicillin 252–58
Penicillium chrysogenum 257
Penicillium notatum 253, 256
Pepper, John Henry 236–39
Pepper's Ghost 236–39
Pericardium–6 142
Permanent dye 176, 178
Pesticides 80–82, 84, 85
Pfizer 252
Phaeomelanin 176
Phenylalanine 43, 73
Phenylketonurea (PKU) 43, 73
Phipps, James 247
Phonograph 234
Phosphoric acid 174
Phosphorus 196, 198, 200, 209, 234
Physostigmine 162, 163
Phytopharm 94
Phytosterols 35
Pica 192
Pig manure 89
Pincus, Gregory 232
Pneumococcus 254
Poirot, Hercule 189
Polyurethane 219
Polyacrylamide 25, 26
Polyester 157, 158
Polyethylene glycol 159
Polyphenols 21, 72
Polystyrene 241
Pomegranate 21–23, 24
Pork production 87
Post-polio myalgia 269
Potash 198
Potassium 196, 198, 200, 212
Potassium chlorate 204–6
Potassium nitrate 206
Prayer 119–22
Priestley, Joseph 205, 206
Progesterone 227–32

Prohibition 182, 183
Propylene glycol 168, 169, 171
Propylene glycol alginate 220
Propylene glycol n-propyl ether 171
Prosperm 86, 88
Prostate cancer 18, 19
Proteins 218
Prozac 154
Pterostilbene 24
Pyramid Power 270
Pyrazine 259
Pyrethrum 84, 85
Pyroglycerine 243
Pyrotrons 264

Radar 258
Ragwort 207
Rauscher, Frances 127
Raytheon Company 258
Réamur, René de 222, 223
Reactive arthritis 57
Recombinant DNA technology 77
Reefer Madness 145
Reeser, Mary 262
Resorcinols 177
Resveratrol 40, 59–61, 76
Retallack, Dorothy 126
Retinal 76
Retinol 76
Rheumatoid arthritis 65
Ricin 136, 137
Ridley, Frederick 253
Rio Hair Naturalizer 66
Rock, John 232
Rockefeller Foundation 257
Rosenkrantz, George 229–31
Rosin 225
Rotenone 85
Rottnest Island 192
Ruzicka, Leopold 229, 230

Saccharin 43
Sacks, Oliver 134
Sainte-Claire Deville, Henri 212
Salmonella 57–58
Saltpeter 198, 206
San Bushmen 93, 94, 96
Sanger, Margaret 232
Sarsaparilla root 228, 230
Sarsaponegin 228
Sativex 149
Saturated fat 30, 31
Saunders, Frederick 240
Schaffer, Jacob 223
Scheele, Carl Wilhelm 173, 206
Schizophrenia 75
Schonbein, Friedrich 243
Schueller, Eugene 175, 176
Scurvy 110
Searle 232
Selenium 88, 101
Semi-permanent dyes 177, 179
Sendivogius, Michael 206
Serotonin 103, 152
Sertraline (Zoloft) 154
Sham acupuncture 141
Shaw, Gordon 126–27
Shellac 52, 53
Silicon dioxide 173
Silicones 219
Silver nitrate 209
Silver sulfide 216
Simethicone 220
Sinclair, David 59, 60
Sir2 59
Sirtuin 59, 60
Sizing 225, 226
Smallpox 245–48
Smoked meat 69–73
Sobrero, Ascanio 243
Sodium bicarbonate 219
Sodium chlorate 207
Sodium hydroxide 224

Sodium nitrate 70, 197, 199
Sodium stearate 235
Sodium sulfide 224
Solar flares 261
Soluble fiber 19
Somatids 112
Somlo, Emeric 229, 230
Sorbitol 54
Splenda 42, 43
Spontaneous human combustion
   261–65
SSRIS 154
Stanozolol 186
Starch 225
Stearic acid 235
Steroids 227, 228, 230–33
Stott, Daniel 190
Stradivari, Antonio 239
Stradivarius 239–41
Straphylococci 253
Stratful, Dan 252
Streptococcus 246
Streptococcus mutans 54
Streptomyces murinus 38
Stroke 60
Strychnine 188–91
Strychnine bromide 188–89
Styrene 27
Sucralose 43–45
Sucrose 37, 43
Sugar 36, 37, 38, 45, 54
Sugar Association 37
Suicide 80
Sulfite 224
Sulfur 180
Sulfuric acid 123, 243
Sun flares 264
Swiffer WetJet 170–71
Syncrometer 117, 118
Syngenta Seeds 77, 78
Syntex 229–33
Synthesis gas 184

Taiwan 132
Tatar, Mark 59
Tea 72
Teleportation 266
Temporary dyes 177–79
Testosterone 19, 126, 158
Texas City 202–3
The Pill 226–33
Thermometer 194
Thiazole 259
Thiosulfinates 214, 215
Thompson, Dr. Lillian 19
Tin 248–52
Tin can 250
Tin disease 249
Tin oxide 251
Titanium dioxide 225, 226
Toluene 159, 160
Toluene-2,5-diamine sulfate 177
Tomatoes 83
Tornquist, Margareta 26
Tourette, Gilles de la 153
Tourette's syndrome 153, 154
"Trans" 30, 32
Trans fat(s) 29–32
Triclosan 128
Trudeau, Kevin 274–76
Turmeric 12
Twain, Mark 11
Type II diabetes 40, 41, 65

Ulcerative colitis 51
Unilever 95
Unsaturated 30
Upjohn Company 231
Urea 199
Urea-formaldehyde 200
Urethane 27

Vaccination 245, 247–48
Van Leeuwenhoek, Antonie 123
Vancomycin 124
Verne, Jules 211
Viagra 36
Vinson, Joe, 65
Violin 239–42
Vitamin A 49, 76–78, 101
Vitamin C 86, 101, 106, 108–12
Vitamin E 13, 88, 101–8
Vitamin supplement 52, 100
Von Liebig, Justus 198, 261, 262

Walford, Dr. Roy 96, 98, 99
Walters, Larry 210
Watson-Watt, Robert 258
Weed killer 207
Weil, Dr. Andrew 69
Weinstock, Dr. Joel 51
Wheatgrass 45–49
Whey protein 50, 51
Wigmore, Ann 46, 47, 49
Williams, Ted 165, 168
Wilson, Simon 137, 138
Wirth, Daniel 122
Wohler, Friedrich 211
Women's Health Study 107
Worms 51
Wright brothers 213

714X 112–15
Xylitol 54, 55

Zemel, Dr. Michael 90–93
Zeolites 89
Zero Point Energy 270, 271
Zoloft 154

About the Author

Joe Schwarcz is a professor of Chemistry and the Director of the Office for Science and Society at McGill University in Montreal, Canada. He hosts a popular weekly phone-in radio show, makes numerous television appearances, frequently gives entertaining and educational public lectures, and writes a column for the Montreal *Gazette*. He has received many honors, including the American Chemical Society's prestigious Grady-Stack Award for Interpreting Chemistry for the Public. In 2005 he received the Royal Canadian Institute's Sanford Fleming medal in recognition of his outstanding achievements in the promotion of knowledge and the understanding of science.